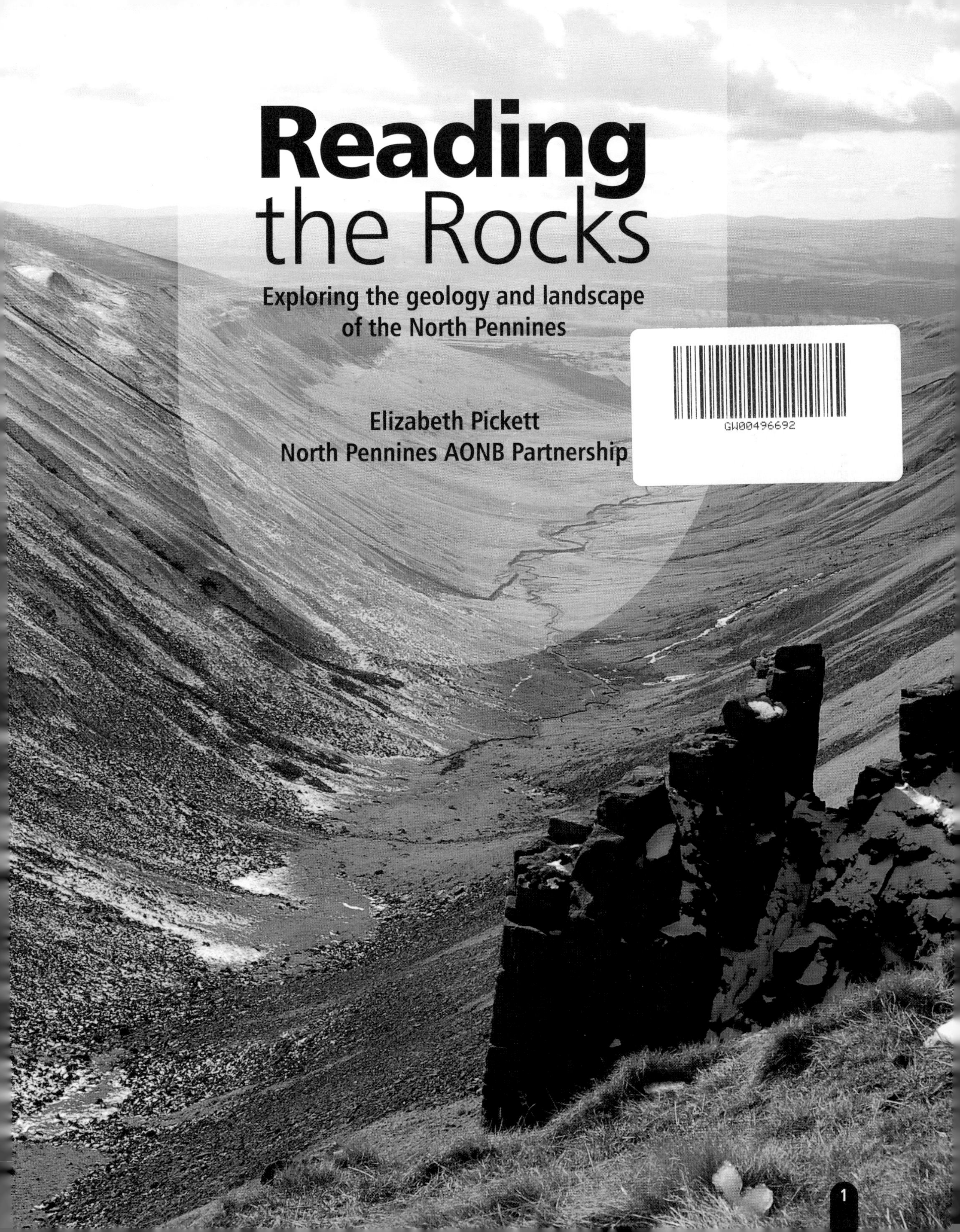

# Reading the Rocks

## Exploring the geology and landscape of the North Pennines

Elizabeth Pickett
North Pennines AONB Partnership

## Acknowledgements

This book has been produced by the North Pennines AONB Partnership with funding from the Heritage Lottery Fund, Natural England and the Friends of the North Pennines.

Written and illustrated by:
Elizabeth Pickett (Geologist, North Pennines AONB Partnership)

Designed by:
GDA (www.gda-design.co.uk)

Produced by:

North Pennines AONB Partnership
Weardale Business Centre
The Old Co-op Building
1 Martin Street, Stanhope
Bishop Auckland
County Durham DL13 2UY

Supported by:

The author's understanding of North Pennine geology has benefited over the years from discussions with geologists and others with local expertise, especially Stu Clarke, Ian Forbes, David Lawrence and Brian Young. The author also acknowledges the information contained in numerous publications, including those on page 48 (and references therein). We are particularly grateful to David Lawrence of the British Geological Survey for constructive comments on a draft of the book.

We thank everyone who generously provided photographs and permission to use them; acknowledgement for individual images is given here. All hand-painted illustrations and uncredited photographs are by Elizabeth Pickett.

This book is printed on: 350gsm (cover) and 170gsm (inside pages) 100% recycled FSC approved Cocoon 100% Silk.

Front cover: *Greenleycleugh Crags, Kevelin Moor, West Allen Valley © David Taylor Photography*

Title page: *High Cup Gill, near Dufton © Natural England/ Steve Westwood*

**Picture credits**
(**Key:** t: top; r: right; l: left; c: centre; b: bottom; NPAP: North Pennines AONB Partnership; BGS: British Geological Survey; NERC: Natural Environment Research Council; USGS: US Geological Survey)

**p3** © Natural England/Charlie Hedley; **p4** © Shropshire Hills AONB/Phil Holden (tc) © English Riviera Tourism Company (c); **p6** Elizabeth Pickett © NERC (c); **p7** Simplified geology map based upon an image derived from BGS digital data by permission of the British Geological Survey. Base map © Crown Copyright. All rights reserved. Durham County Council LA100049055. 2011 (c); **p8** © Robert Krimmel/USGS (bl); **p10** Elizabeth Pickett © NERC (tl); © NPAP/Shane Harris (br); **p11** © USGS (tl); © Natural England/Charlie Hedley (b); **p12** Elizabeth Pickett © NERC (c); © K.L. McKee/USGS (b); **p13** © NPAP/K. Gibson (cl & bl); **p19** Elizabeth Pickett © NERC (tl); **p20** Courtesy of William Stobbs (cl); © Aggregate Industries Ltd (Heights Quarry) (b); **p21** Courtesy of Judith Bainbridge (tl); © NPAP/Neil Diment (tc); © Natural England/Charlie Hedley (tr); **p22** © Ladycross Quarry (c); **p23** © NPAP/Elfie Waren (tr); © NPAP/Chris Woodley-Stewart (c); **p24** © John O'Reilly (tr); **p25** © Natural England/Charlie Hedley (b); **p26** Reproduced by permission of the British Geological Survey © NERC. All rights reserved. IPR/142-67C (cr); © NPAP/Abi Wylde (bl); **p28** © Beamish Museum Ltd (tr); **p29** © NPAP/Neil Diment (tl); © Andrew Hayward (b); **p30** © Jesse Fisher/UK Mining Ventures (tr); Elizabeth Pickett © NERC (bl); **p31** © Killhope Lead Mining Museum (tr); © Jesse Fisher/UK Mining Ventures (cl, cr & b); **p32** © Beamish Museum Ltd (cl & cr); **p33** © Natural England/ Charlie Hedley (tl); **p35** © Peter Jackson (tl); **p37** © Penrith & Eden Museum (br); **p38** © NPAP/Shane Harris (l); © Kent Downs AONB Unit (br); **p39** © Olikristinn (http://creative commons.org/licenses/by/3.0/deed.en) (b); **p40** © Natural England/Steve Westwood (cr); © Mike Embree/US National Science Foundation (b); **p42** © Simon Ledingham (cl); © D.J.A. Evans (b); **p44** © Natural England/Charlie Hedley (tr); **p45** © NPAP/Paul Leadbitter (bl); **p46** © NPAP/Paul Frodsham (tr); **p47** © NPAP/Lesley Silvera (tl); © NPAP/Neil Diment (cl); © NPAP/Charlie Hedley (b).

# Welcome to the North Pennines

Welcome to the North Pennines Area of Outstanding Natural Beauty (AONB) and European & Global Geopark. This is a special place of high moorland, wild fells, waterfalls, green dales, stone walls, scattered settlements, fantastic wildlife and a rich industrial heritage.

This unique landscape is the result of millions of years of Earth history and a few thousand years of human activity. The character of the North Pennines has its foundation in the underlying rocks and the geological processes which have shaped the fells and dales. These have influenced the plants and animals that live here, and the ways in which people have used the landscape through the ages.

▼ *The moors above Ireshopeburn, Upper Weardale*

## Contents

# A special landscape

Because of its stunning landscape the North Pennines is an Area of Outstanding Natural Beauty (AONB). It is also a UNESCO-endorsed European and Global Geopark in recognition of its superb geological heritage.

## An Area of Outstanding Natural Beauty

Areas of Outstanding Natural Beauty (AONBs), along with National Parks, represent some of our finest countryside and are within a worldwide category of protected landscapes. The designation of the North Pennines AONB was confirmed in 1988. It includes parts of the counties of Durham, Northumberland and Cumbria, and at nearly 2,000km$^2$ it is the second largest of the 38 AONBs in England and Wales. Visit: **www.northpennines.org.uk** and **www.landscapesforlife.org.uk**

▲ *The Shropshire Hills, another AONB with a rich geological heritage*

▼ *The English Riviera Global Geopark*

## A European and Global Geopark

The North Pennines AONB is also Britain's first European Geopark. It was awarded this UNESCO-endorsed status in 2003, and in 2004 became a founding member of the Global Geoparks Network. Geoparks are places with outstanding geology where special effort is made to make the most of Earth heritage through interpretation, education, conservation and tourism. There are several other Geoparks in the UK and a growing number around Europe and the rest of the world. Visit: **www.europeangeoparks.org**

## Exploring the North Pennines

In each section in this book there are suggestions of interesting places to visit. These are just a small sample of what the North Pennines has to offer; if you get out and about you'll discover many more. The grid references are for the sites themselves rather than viewpoints. Where a location covers an area or is a linear feature, the grid reference is for a roughly central or other obvious point.

Unless otherwise indicated, the locations are accessible or visible from public roads, public rights of way or permissive paths, or are on access land (shaded yellow on new OS Explorer maps).

Please follow the Countryside Code and Moorland Visitor's Code (for more information on these and access land visit **www.northpennines.org.uk**). Please be particularly careful around quarries (both disused and working) and old mine sites. These can be dangerous places so please view from a safe distance.

# A moving story

The landscape of the North Pennines tells a story that began almost 500 million years ago – at a time when the Earth was a very different world and Britain as we know it did not exist.

## Dynamic Earth

Over millions of years, the piece of the Earth's crust containing what is now the North Pennines has travelled vast distances over the surface of the globe. This is because of plate tectonics, the process by which the plates that make up the outer layer of the Earth are constantly on the move. These plates, which carry the continents and oceans, pull apart, collide and slide past each other. Where they pull apart, oceans are created and where they collide, oceans close and mountain ranges rise. The plates move a few centimetres a year – about the speed our fingernails grow. This may not sound much but over millions of years plates move vast distances. Because of plate tectonics the North Pennines has been on a remarkable journey, one that is recorded in the area's rocks, fossils and landscapes.

▼ *The outer layer of the Earth is constantly changing, as plates carrying the continents and oceans pull apart and collide*

## Stories in stone

During this incredible journey the North Pennines has been shaped by many different geological processes, environments and climates. The rocks that form the fells and dales of the North Pennines tell of this journey. By reading the landscape and spotting clues in the rocks, we can discover a fascinating story – of a deep ocean and violent volcanoes, colliding continents and molten rock, tropical seas and lush rainforests, hot water and minerals, desert dunes and vast ice sheets. In the last few thousand years – just the blink of an eye in geological terms – North Pennine people have further shaped the landscape with settlements, farms, quarries and mines. And it is still evolving – through a combination of natural processes and human activity.

**We hope this book will encourage you to discover and explore the North Pennine landscape and find out more about the fascinating story it tells.**

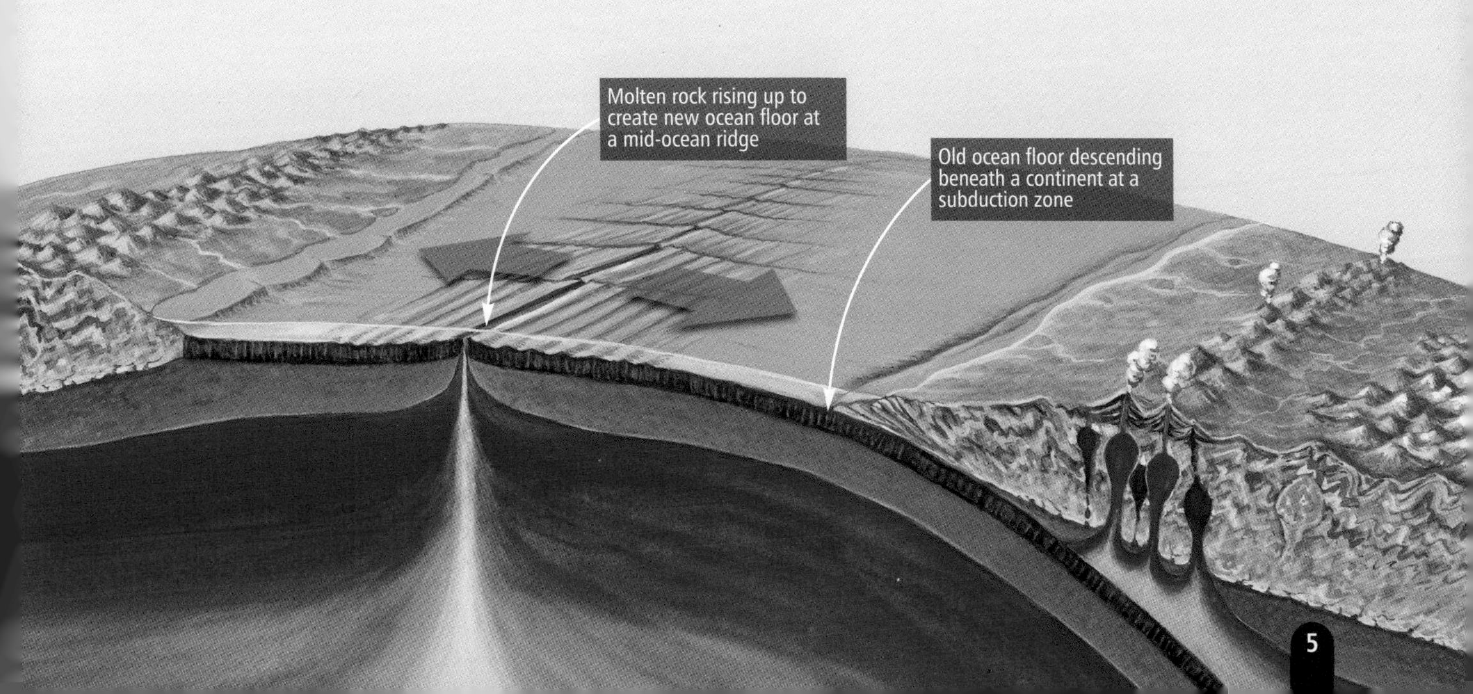

# Foundations of the landscape

Rocks are the basic building blocks of the landscape and influence every aspect of the countryside. The many types of rock that make up the North Pennines can be seen on the geological map and cross-section. They formed in different ways and at different times in Earth history – as you'll discover in this book.

## A slice through the North Pennines

This imaginary slice through the North Pennine escarpment, from the Eden Valley to the Cross Fell range, shows some of the main features of the landscape and how they relate to the underlying rocks. The labels indicate the main rock units – see the timeline below for when these formed.

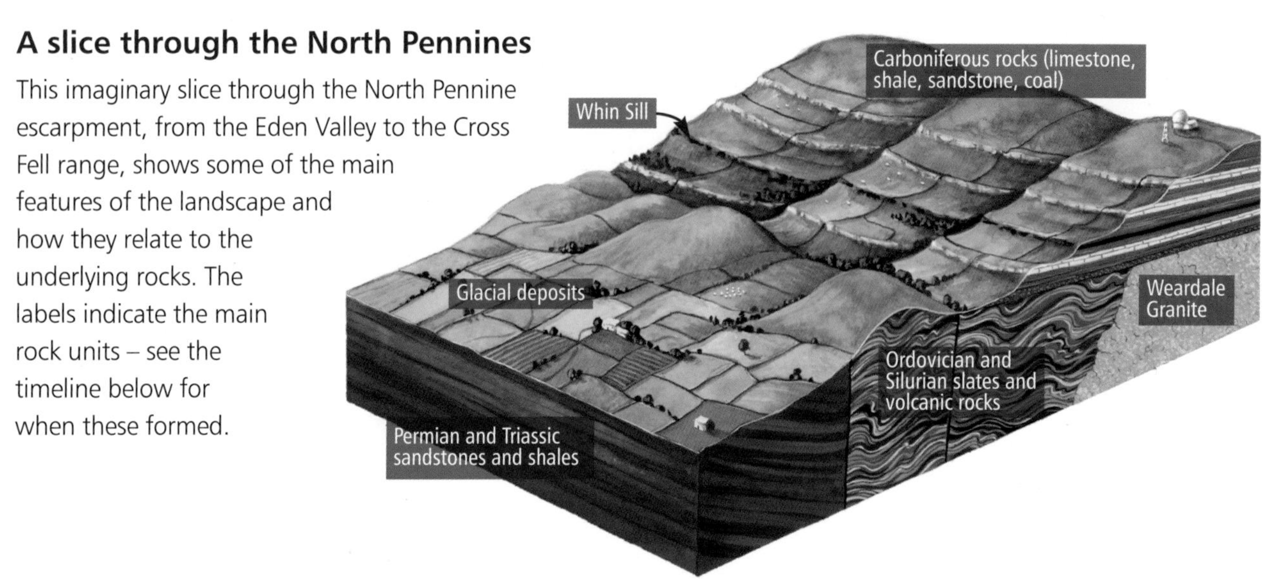


## Journey through time

This timeline for the North Pennines extends from the birth of the Earth to the present day. It is divided into chunks of time known as periods. The coloured time intervals indicate the periods for which there is evidence in the North Pennines, and there are brief descriptions of the geological events that have created today's landscape – read on to find out more…

| Millions of years ago | 4,600 to 542 | 542 to 488 | 488 to 444 | 444 to 416 | 416 to 359 | 359 to 299 |
|---|---|---|---|---|---|---|
| **Period** | Precambrian | Cambrian | Ordovician | Silurian | Devonian | Carboniferous |
| | A vast period of time during which the first life on Earth emerged. The oldest rocks in Britain – the Lewisian Gneisses of NW Scotland – are about 3,000 million years old | The Iapetus Ocean separated the continents of Laurentia (including Scotland) and Avalonia (including England) | Mud, sand and volcanic ash accumulated in the Iapetus Ocean as it started to close | The Iapetus Ocean closed and the continents of Laurentia and Avalonia collided creating the Caledonian Mountains | The Weardale Granite was injected deep into the roots of the Caledonian Mountains, which eroded rapidly in a hot climate | The North Pennines lay at the equator and was covered by tropical seas, deltas and rainforests. Most North Pennine rocks formed at this time |

## A map of the rocks

This simplified geological map of the North Pennines shows the main rock units which lie at or near the Earth's surface. Different rocks may exist deeper underground, as shown on the cross-section.

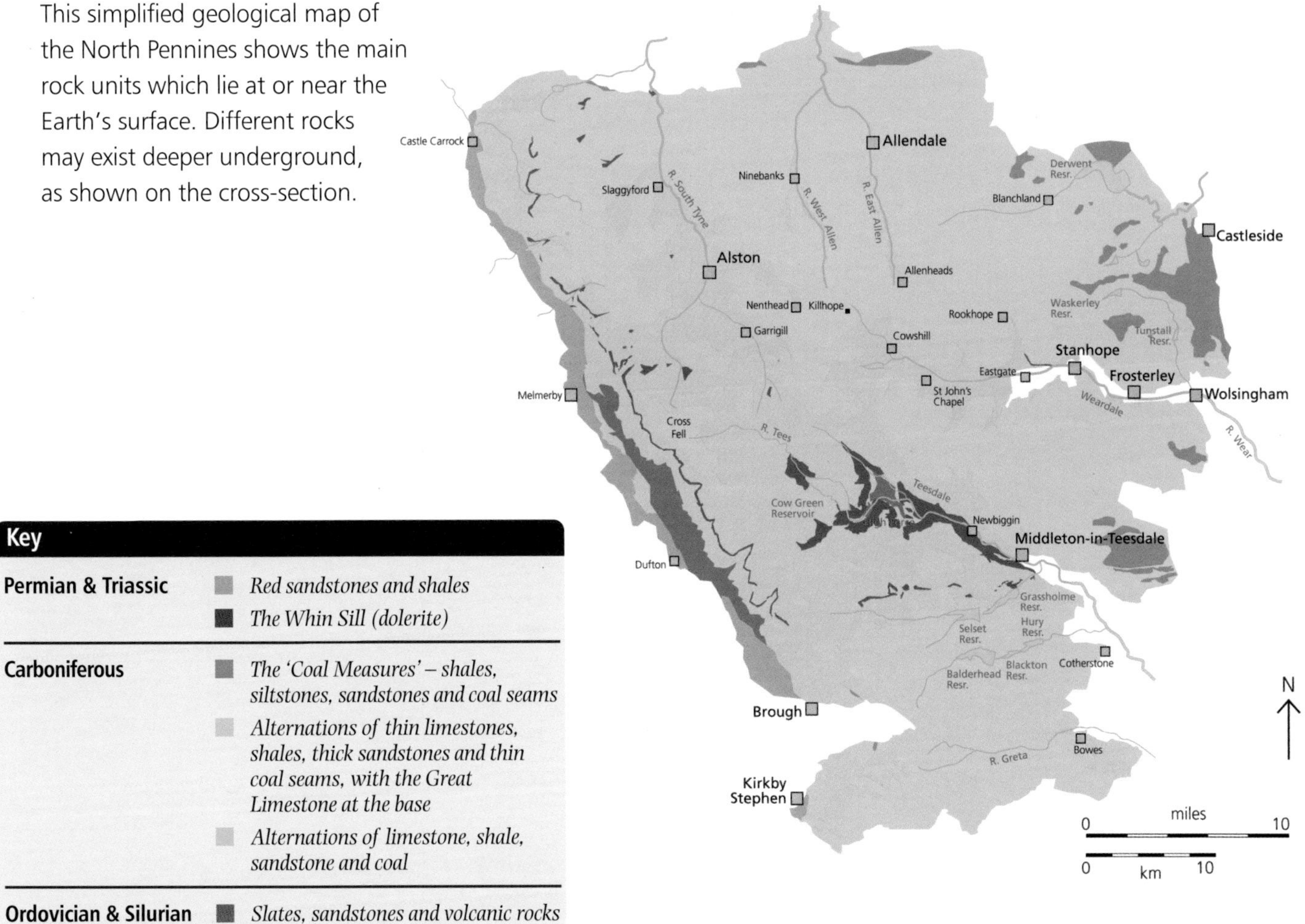


| 299 to 251 | 251 to 200 | 200 to 145 | 145 to 65 | 65 to 23 | 23 to 2.6 | 2.6 to present |
|---|---|---|---|---|---|---|
| **Permian** | **Triassic** | **Jurassic** | **Cretaceous** | **Palaeogene** | **Neogene** | **Quaternary** The Ice ages |
| Early on, the Whin Sill formed from molten rock, followed by the formation of mineral deposits. The North Pennines was hot and arid and covered in desert dunes and salty lakes | Desert conditions continued from the Permian Period. Great rivers flowed over desert plains | Shallow seas covered much of Britain, but deposits from this time later eroded off the North Pennines | Most of Britain lay under a warm sea and was covered in thick chalk. This later eroded off most of northern Britain | The North Atlantic Ocean started to open, accompanied by volcanic activity in western Scotland. Intrusions of molten rock stretched as far as the North Pennines | The North Pennines was uplifted and eroded in warm, humid conditions | Glacial periods alternated with warmer periods. Ice covered the North Pennines several times |

# Deep roots

The oldest rocks in the North Pennines are slates and volcanic rocks, which formed between 480 and 420 million years ago, in the Ordovician and Silurian periods of Earth history. These rocks tell a story of a long-vanished ocean, violent volcanoes and colliding continents. They are mostly buried beneath younger rocks and form the deep roots of the North Pennine landscape.

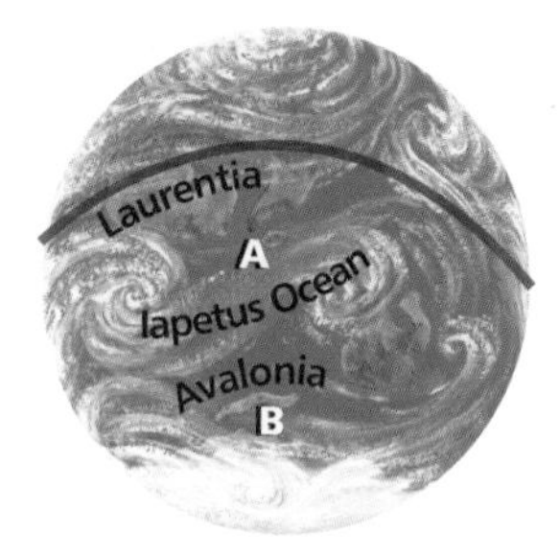


▲ *How the world may have looked 500 million years ago. 'A' marks the position of Scotland, and 'B' that of England and Wales*

## Ocean of time

About 500 million years ago a wide ocean, known as the Iapetus Ocean, lay between the pieces of the Earth's crust which would eventually come together to form Britain. This ocean, which once separated England and Scotland, was a few thousand kilometres across and lay south of the equator. On its northern margin was the continent of Laurentia, which contained Scotland, the north of Ireland, Greenland and North America. South of the ocean lay Avalonia, a small continent which contained England, Wales and the south of Ireland.

## Blast from the past

About 470 million years ago the continents of Laurentia and Avalonia began to drift together, and the Iapetus Ocean started to narrow. The ocean floor sank into the Earth in a process known as subduction, bringing the two continents inexorably closer. Volcanoes at the edge of the closing ocean spewed out ash and lava, which became interleaved with ocean-floor mud, silt and sand.

▼ *The volcanic rocks found along the North Pennine escarpment were formed in violent volcanic eruptions similar to that of Mount St Helens, USA, in 1980*

▼ *This pale streaky rock from scree on Dufton Pike is a volcanic rock known as ash-flow tuff*

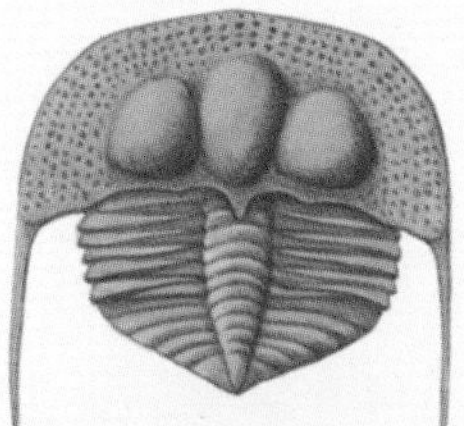

▲ *This trilobite species lived in the Iapetus Ocean and was about 3cm across*

## Strange lives

Land plants and animals had yet to evolve, but the world's oceans were full of life. Trilobites crawled on the ocean floor and stick-like colonies of tiny creatures, called graptolites, drifted in the water. When they died they were buried in the mud and silt at the bottom of the ocean. These sediments eventually hardened into mudstone and siltstone, preserving the long-extinct creatures of the Iapetus Ocean as fossils.

## United Kingdom

Laurentia and Avalonia finally collided about 420 million years ago and the mighty Caledonian Mountain range rose up along the collision zone. Remnants of these mountains can be traced today from North America to Scandinavia. The collision brought the two halves of Britain together along a weld that is now deep in the Earth's crust below the English–Scottish border. Remnants of this ancient mountain chain can be seen in the Scottish Highlands, from where they take their name.

▲ *Outcrop of slates in Gasdale, near Murton*

## Heat and pressure

During the collision, the mudstone, siltstone, sandstone and volcanic rock that formed on the Iapetus Ocean floor were caught between the two continents. They were squashed, heated and crumpled to become hard, slaty rocks. These rocks form the Lake District fells and the basement of the North Pennines. Although they are mostly buried beneath layers of younger rock, you can see them in a few places – along the foot of the North Pennine escarpment and in one part of Upper Teesdale.

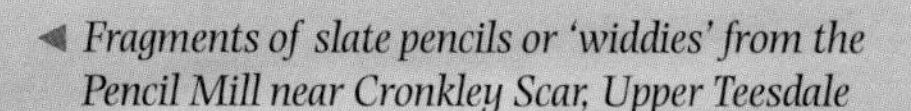

◄ *Fragments of slate pencils or 'widdies' from the Pencil Mill near Cronkley Scar, Upper Teesdale*

▲ *Dufton Pike from Keisley*

## WHERE TO SEE

- **Knock Pike [NY 686 282] and Dufton Pike [NY 700 266]**
  These distinctive conical hills along the foot of the North Pennine escarpment are made of slates and volcanic rocks, which can be seen in outcrops, quarries and walls.

- **Pencil Mill, Upper Teesdale [NY 848 296]**
  This small, disused quarry is in greenish slates beside the River Tees near Widdybank Farm. In the mid 1800s this was the site of a small factory making slate pencils or 'widdies'.

▼ *Looking south-east along the North Pennine escarpment from Knock Pike. Dufton Pike is in the middle distance and in the far distance is Murton Pike*

# Granite and deserts

After the Iapetus Ocean closed and the Caledonian Mountains were created, molten rock rose up and cooled to form large masses of granite deep within the collision zone. Above ground, Britain was south of the equator and part of a vast red desert continent. This was the Devonian Period, which lasted from around 420 to 360 million years ago.

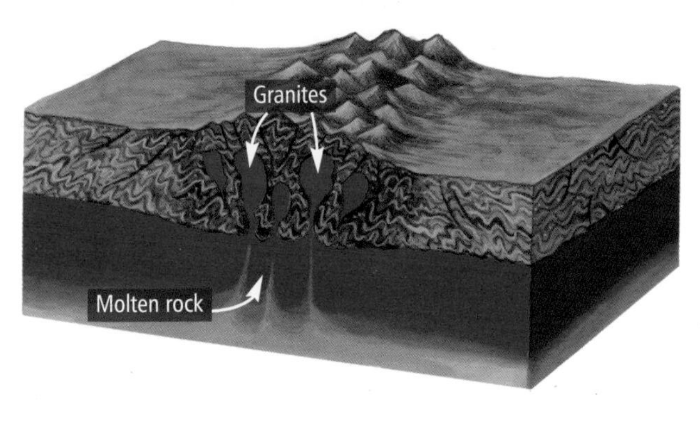


▲ *Molten rock rising up into the Caledonian Mountains and cooling to form granite masses*

## The Weardale Granite

In the final stages of collision, about 400 million years ago, molten rock rose up into the roots of the Caledonian Mountains. It cooled and solidified in the Earth's crust to form huge granite bodies. One of these is the Weardale Granite, which underlies the North Pennines.

## Buoyed up by granite

Although hidden deep underground, the Weardale Granite is a fundamental geological feature. Granite is less dense than most other rocks in the Earth's crust and is relatively buoyant. Because of this, the area above the granite – much of the North Pennines – has remained higher than surrounding areas for many millions of years. The granite also played an important role in the formation of the area's mineral deposits, as we shall see later.

▼ *A piece of the Weardale Granite, from a borehole drilled at Rookhope by Durham University researchers in 1960-61. This piece is 13cm across*

▲ *The site of the Rookhope borehole. Granite was reached 390m below the surface*

▲ *The North Pennines 400 million years ago would have been similar to present-day Arizona, USA*

## Hot red land

About 400 million years ago, Britain's climate was hot and semi-arid – similar to that of Jordan or Arizona today. Simple plants had only just begun to colonise the land. The Caledonian Mountains eroded rapidly, their bare slopes worn down by water and wind. Flash floods carried boulders and sand down to desert plains. These sediments hardened into red rock, traditionally known as the Old Red Sandstone, which is found today in various parts of Britain.

▲ *Red conglomerate near Melmerby*

## Rocks within rocks

Gravel and boulders accumulated amongst the eroding mountains, eventually hardening to conglomerate – a rock made of fragments of older rocks. A few small exposures of red conglomerate, believed to be from this time, occur along the North Pennine escarpment. This conglomerate contains chunks of slate and volcanic rock, probably once part of the ancient Caledonian Mountains.

## Building blocks

The end of the Devonian Period heralded the start of a period of stretching and rifting of the Earth's crust in northern Britain. Fractures, or faults, developed around the area that was underpinned by the Weardale Granite. Geologists know this upstanding block of crust, which roughly corresponds to the North Pennines, as the Alston Block. The faults that bound the North Pennines influence the landscape to this day; the dramatic North Pennine escarpment follows the line of the Pennine Fault System.

**WHERE TO SEE**

▲ *Harehope Quarry, Frosterley, where a piece of Weardale Granite is set into this sculpture*

You can see samples of Weardale Granite at: Killhope, the North of England Lead Mining Museum, Harehope Quarry and in the geology display at Langdon Beck Hotel in Upper Teesdale. All these samples came from a borehole drilled at Rookhope in 1960–61.

**Melmerby track [NY 628 370]**
Red conglomerate, thought to be of Devonian age, is exposed beside a footpath about 1km east of Melmerby.

▼ *The North Pennine escarpment, which marks the western edge of the Alston Block, viewed from the Eden Valley*

# Carboniferous world

Most North Pennine rocks date from the Carboniferous Period, which lasted from about 360 to 300 million years ago. The landmass that contained Britain had drifted north to the equator and basked in a hot and humid climate, similar to that of the Bahamas today. Sea level was continually changing, and shallow tropical seas, vast river deltas and lush swampy forests periodically covered northern England.

## Seas, swamps and cycles

By 360 million years ago the Caledonian Mountains had worn down to their roots and the landmass that contained Britain had drifted close to the equator. We were about to experience over 50 million years of tropical climate and changing sea levels.

Warm, shallow seas flooded over northern England. Sea creatures flourished and their remains accumulated as limy ooze on the sea floor. Vast rivers drained into the sea from the north, bringing with them sand, silt and mud. The rivers built up large deltas on which swampy rainforests grew.

▲ *About 350 million years ago the North Pennines was astride the equator*

In time, the limy ooze became limestone, the mud and sand became shale and sandstone, and the forests turned to coal. Periodically the sea rose, flooding the deltas and depositing limestone again. This cycle happened many times, building up repeating layers of limestone, shale, sandstone and coal, known as cyclothems. Eventually, the deltas became well established and flooding by the sea was less frequent. Great forests flourished on the deltas, eventually becoming the coal seams of the Durham Coalfield to the east.

◄ *This aerial view of the Mississippi River Delta and the distant Gulf of Mexico gives an idea of how the North Pennines might have looked in Carboniferous times*

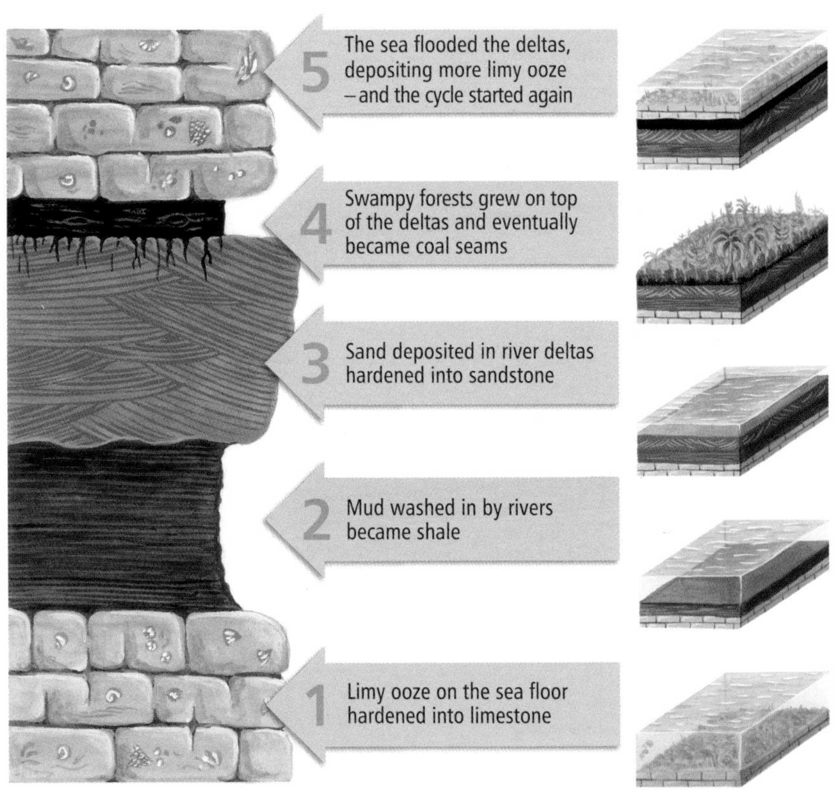


▲ *How North Pennine cyclothems formed*

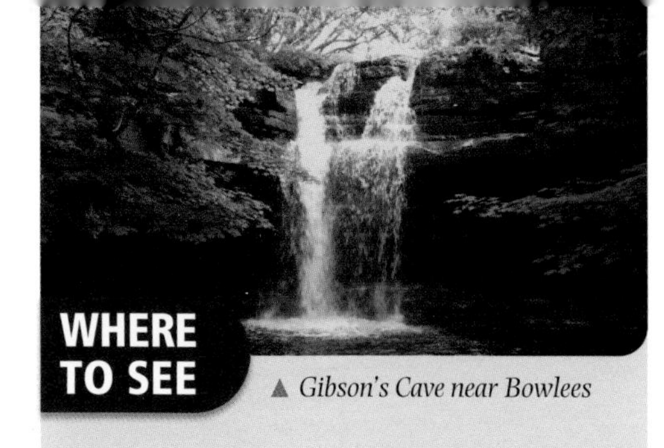

▲ *Gibson's Cave near Bowlees*

**WHERE TO SEE**

**Ashgill Force [NY 759 405] and Summerhill Force at Gibson's Cave [NY 909 286]**
These dramatic waterfalls tumble over resistant limestone layers which overlie softer, more easily eroded shales and sandstones.

**Blagill [NY 740 475]**
There are very clear terraced hillsides around Blagill in the Nent Valley, best seen from the A698, just east of Alston.

You can also see good terraced hillsides in parts of Weardale, Teesdale and the Allen Valleys.

▲ *Terraced hillsides around Blagill in the Nent Valley*

## Sea change

The changing sea levels of the Carboniferous may be partly related to a series of major glaciations at the South Pole. During the glaciations water was locked up in vast ice sheets, causing sea level to fall globally. When the ice melted, sea level rose. More locally, faults also probably played a role by causing bits of the Earth's crust to move up or down relative to sea level.

## Layered landscape

The North Pennine landscape looks the way it does because of cyclothems. Limestone and sandstone are hard rocks which are resistant to erosion, whereas the softer shales wear away more easily. This contrast has produced the area's distinctive terraced hillsides, lines of crags and flat-topped hills. Some North Pennine waterfalls are good places to see sections through cyclothems. Hard layers of limestone commonly form resistant lips, while underlying softer shales and thin sandstones are more easily worn away by the water.

◀ *Ashgill Force in the South Tyne Valley, showing a typical sequence of shale, sandstone and limestone*

# Coral seas

The tropical seas of the Carboniferous Period were full of life. Sea creatures, such as corals, crinoids, sponges and shellfish, flourished in the warm, sunlit waters. When they died their remains accumulated as limy ooze and shelly fragments on the sea floor. These deposits – made of calcium carbonate like modern shells – eventually hardened into limestone. We know about the creatures that lived in these ancient seas from their fossilised remains, now entombed in limestone.

## Crinoids

Crinoids, sometimes known as 'sea lilies', were anchored to the sea floor and swayed in the currents, catching particles of food with bony arms. They were not plants, but relatives of starfish and sea urchins. Crinoids had skeletons made of disc-shaped plates. When they died, these plates fell apart and scattered on the sea floor. Crinoids are common fossils in North Pennine limestones – look out for segmented sections of stalk or small white discs with a hole in the middle.

▼ *Fragments of crinoid stalks*

▲ *Fossil corals in polished Frosterley Marble*

## Corals

Many different types of corals lived in these tropical seas. Like modern corals, some lived in colonies, whereas others were solitary. The colonial corals grew in many different shapes, from mounds and platforms to fans and branches. The solitary corals were shaped like horns and cups. The Frosterley Marble of Weardale contains striking white fossils of the horn-shaped solitary coral *Dibunophyllum bipartitum*.

▼ *A Carboniferous tropical sea*

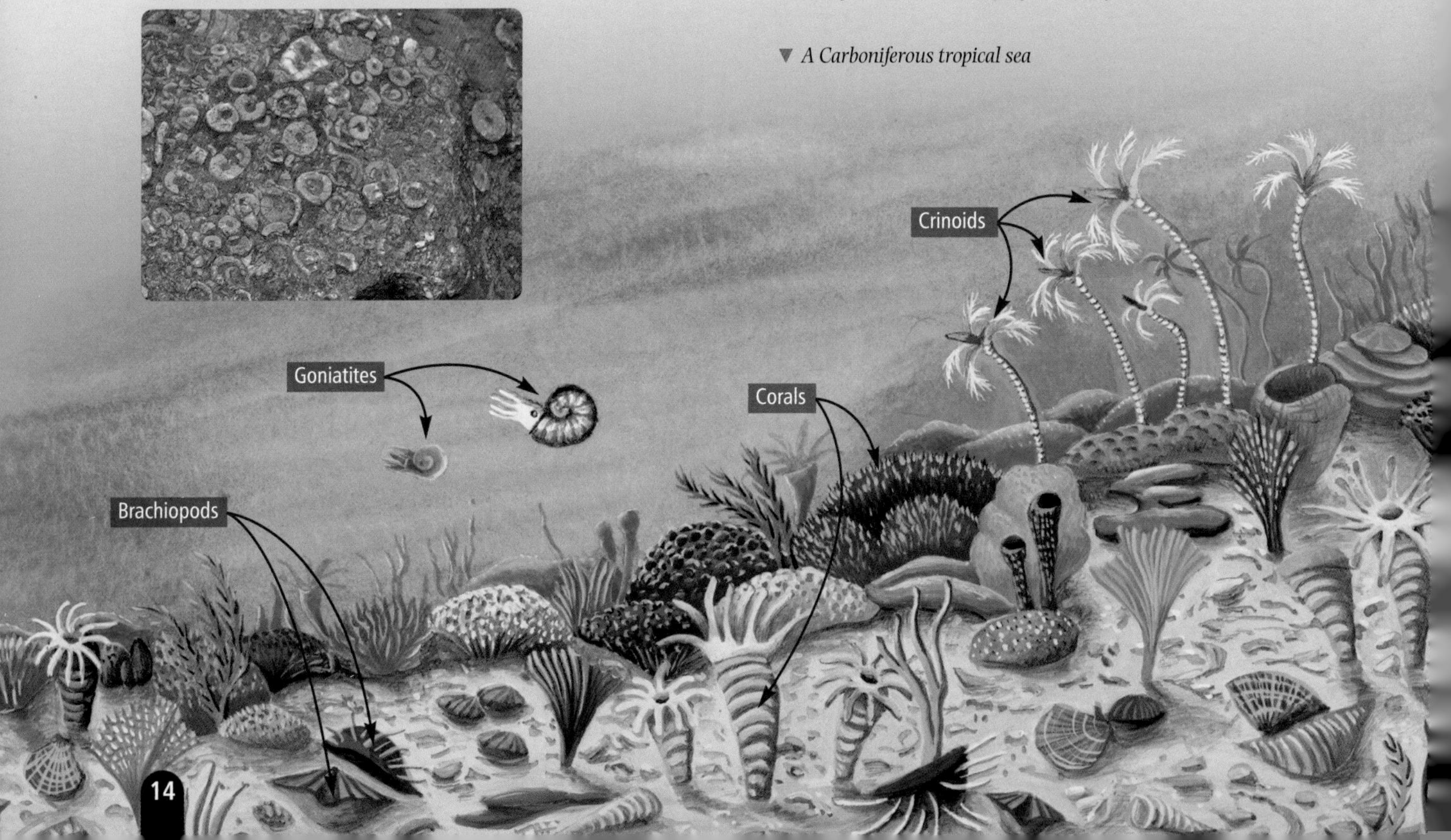

▲ *The curved valves of a brachiopod shell in cross-section, Scoberry Bridge*

## Brachiopods

These filter-feeding shellfish had thick shells and some anchored themselves to the sea floor. They were much more common in the Carboniferous than they are today. Their fossils are common in some North Pennine limestones. Look out for slim, white, curved shapes; these are cross-sections through the shells.

## Molluscs

Many kinds of mollusc lived in the Carboniferous seas. They included bivalves, gastropods, goniatites and nautiloids. Bivalves are filter-feeding shellfish, like modern scallops and mussels. Goniatites had coiled shells and were ancestors of the ammonites of later geological times. They swam in open water, along with nautiloids which had straight shells.

## Sponges

Like modern sponges, these fed by filtering seawater through tiny pores in their outer walls. Some sponges had platy, mat-like shapes; an example is *Chaetetes* which forms a distinctive layer of wavy bands near the base of a thick limestone known as the Great Limestone.

▼ *Layers or 'beds' of limestone in Maize Beck, near High Cup Nick*

## WHERE TO SEE

▲ *Sculpture of Frosterley Marble at the Durham Dales Centre*

- **Durham Dales Centre, Stanhope [NY 996 393]**
  There is a polished sculpture of Frosterley Marble in the courtyard, and a medieval coffin of Frosterley Marble outside nearby St Thomas' Church.

- **Harehope Quarry [NZ 034 361]**
  There is a superb outcrop of Frosterley Marble in Bollihope Burn at this old limestone quarry (now an environmental education centre). The best place to view it is from a footbridge on a permissive path at the west end of the quarry.

- **Scoberry Bridge [NY 910 273]**
  Beside the River Tees at Scoberry Bridge is a fine outcrop of a limestone layer known as the Cockleshell Limestone. It contains impressive fossil brachiopod shells, as well as corals and crinoids.

- **Maize Beck [NY 767 268]**
  Colonial corals are beautifully exposed in limestone in Maize Beck beside the Pennine Way, near High Cup Nick.

You can see different examples of limestone all over the North Pennines. It typically forms crags and quarry faces of layered grey rock.

# Limestone country

Limestone is an unusual rock. Although it is hard and resistant to erosion, it is also slightly soluble in rainwater. This characteristic feature of limestone creates special natural features and distinctive landscapes, known collectively as karst. Where there is limestone in the North Pennines we have some beautiful karst features – from sinkholes and caves to limestone pavements and natural bridges.

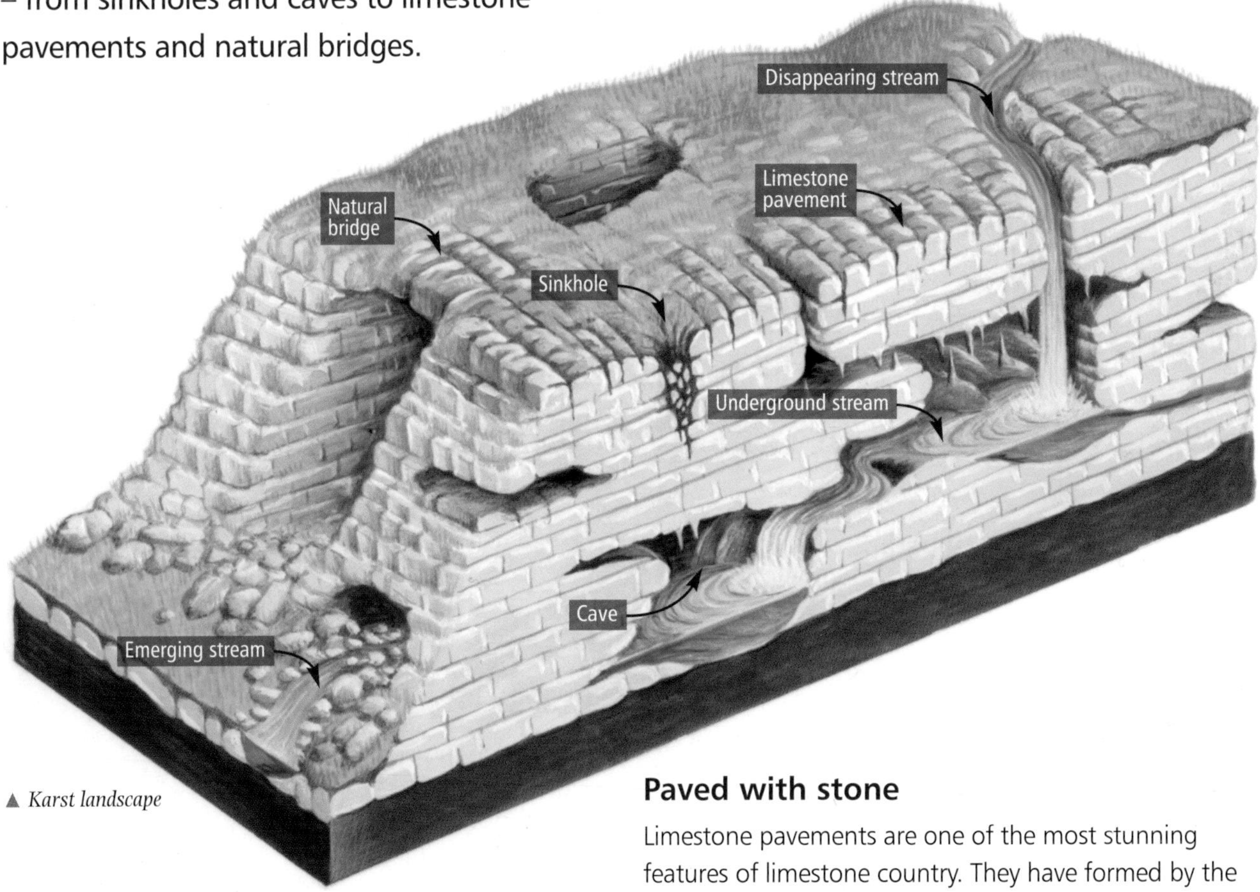

Disappearing stream
Limestone pavement
Natural bridge
Sinkhole
Underground stream
Cave
Emerging stream

▲ *Karst landscape*

## Dissolving rock

Limestone is made of calcium carbonate ($CaCO_3$), which is soluble in acid. Rainwater is slightly acidic because it dissolves carbon dioxide from the air, forming a weak solution of carbonic acid. It also becomes more acidic as it soaks through soil. Rain therefore gradually dissolves limestone, sculpting a range of karst features.

## Paved with stone

Limestone pavements are one of the most stunning features of limestone country. They have formed by the action of rainwater on wide, flat outcrops of limestone, stripped bare in the last ice age. Vertical cracks in the rock widen into fissures known as grykes, which separate blocks or clints. The sheltered grykes provide a habitat for rare and unusual plants.

◄ *Limestone pavement at Stainmore, east of Brough*

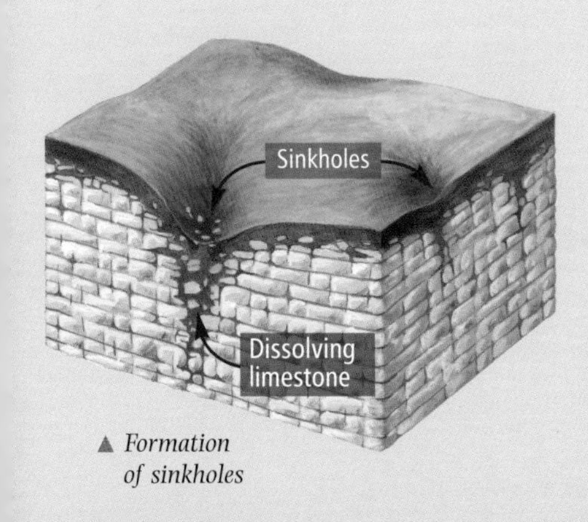

▲ *Formation of sinkholes*

## Sinkholes

Sinkholes, or shakeholes, are the most common karst feature in the North Pennines, and are a telltale sign that limestone lies beneath. They form when rainwater seeps down through limestone, dissolving it and widening cracks. The overlying glacial deposits and soil eventually sink into the cracks, forming distinctive hollows.

## Elusive streams

Streams that disappear are also a feature of limestone country. The water sinks into the limestone and flows underground, leaving dry stream beds which only run after heavy rain. When the water meets an impervious rock such as shale it is forced to come to the surface at a stream 'resurgence', or along a line of springs.

## Hollow hills

North Pennine caves do not match the spectacular cave systems of the Yorkshire Dales, where the limestone is much thicker. However, the Knock Fell Caverns are Britain's finest example of a phreatic maze cave – a cave system formed below the water table. There are also small cave systems in Teesdale and Weardale and near Alston. All these are in the Great Limestone which, at 20m thick, is one of the thickest local limestones.

◄ *Sinkholes on Herdship Fell, Upper Teesdale*

▼ *Limestone country above Helbeck, near Brough*

## WHERE TO SEE

▲ *God's Bridge*

### God's Bridge [NY 957 126]
This remarkable natural bridge in the Great Limestone spans the River Greta near Bowes. It is probably the relic of a collapsed valley floor cave system.

### Helbeck [NY 795 165]
The crags and pavements above Helbeck, near Brough, are in the Great Scar Limestone. Stay in the area of access land east of Warcop MoD Training Area.

### Nateby Common [NY 809 047]
Near the boundary with the Yorkshire Dales National Park, this area of limestone pavement and sinkholes is in the Great Limestone (also known as the Main Limestone in this area).

### Stainmore [NY 853 146]
Limestone pavement developed in the Great Limestone can be seen on the north side of the A66, near Banks Gate.

# Sands of time

Huge rivers drained into the Carboniferous seas from hills to the north. They washed mud, silt and sand into the sea, choking the clear tropical waters and burying the corals and crinoids. Where the rivers dumped their load of sediment, vast deltas built far out into the sea. The soft sands and muds eventually hardened into sandstone and shale, which contain clues to the delta environments and the creatures that lived in them.

## Lasting impressions

The mud, silt and sand of the deltas teemed with worms, molluscs and crustaceans, which burrowed and crawled in the soft sediment. Fossils of the animals themselves are rare, but many local sandstones contain the trails, burrows and feeding traces they left behind. These are known as trace fossils; look out for them in walls, river stones and sandstone paving slabs.

◄ *Fossil worm or mollusc trails on a slab of sandstone at Coalcleugh, West Allen Valley*

## Turned to stone

As the layers of mud and sand were buried beneath more sediments they compacted and hardened. In time, the grains in the sand layers became naturally cemented together, and the sand became solid sandstone. Mud, by contrast, is mainly made of tiny, flaky, clay particles. When mud is buried these flakes become aligned – in the way that sheets of loose paper settle on top of one another. The resulting dark grey rock, known as shale, tends to split along these horizontal planes.

◄ *Sandstone overlying shale in Middlehope Burn, Westgate, Weardale*

▼ *Carboniferous river deltas building out into tropical seas*

## Lines in the sand

If you look at the top surfaces of slabs of local sandstone you may spot fossil ripple marks. These were formed by water flowing over sand in the Carboniferous deltas, in exactly the same way as ripples form today on sandy beaches and riverbeds.

If you see sandstone layers in cross-section, on the face of a crag or quarry, you may notice parallel sloping layers. These are a feature known as cross-bedding which forms when ripples or small dunes migrate along a riverbed. As the sloping layers face downstream, these structures can tell us which way the water currents were flowing over 300 million years ago.

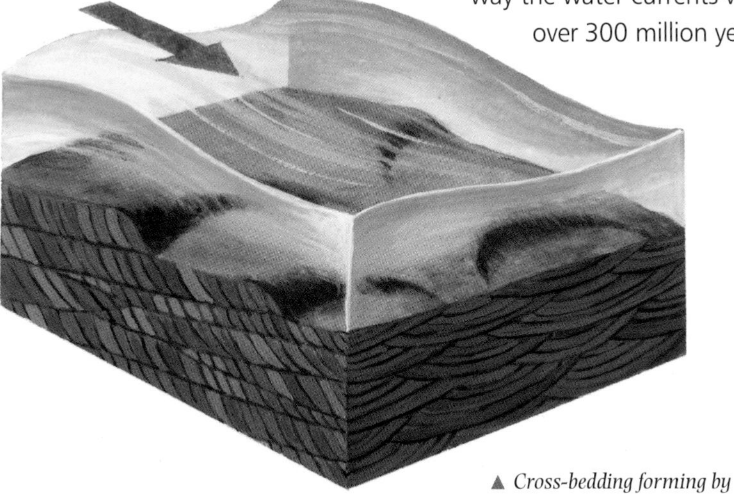

▲ *Cross-bedding forming by migration of sand dunes on a riverbed*

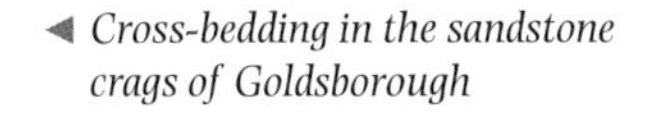

◄ *Cross-bedding in the sandstone crags of Goldsborough*

## Hard rock, soft rock

Sandstone is a hard, resistant rock and commonly forms flat caps on hills and sometimes stands out as low brown crags and terraces. Shale is much softer and wears away more easily. It doesn't form strong landscape features, and good outcrops are relatively rare.

*Shale with thin sandstone beds at Reigill, near Nine Standards Rigg* ►

▲ *Sandstone cliffs at Raven Crag, Allen Banks*

### WHERE TO SEE

- **Allen Banks and Staward Gorge [NY 798 629]**
  Dramatic sandstone cliffs tower over this lovely wooded gorge which is looked after by the National Trust.

- **Shacklesborough [NY 908 170] and Goldsborough [NY 954 177]**
  Gritty sandstone forms these prominent crags on Cotherstone Moor. They show superb examples of cross-bedding.

- **Slitt Wood [NY 906 386]**
  Along the wooded valley of Middlehope Burn in Weardale there are good outcrops of sandstone and shale, as well as limestone and mining remains.

You can see sandstone and shale layers in many quarries and stream cuttings around the North Pennines.

# Quarrying the Carboniferous

Carboniferous rocks have been quarried in the North Pennines for many centuries. The landscape is dotted with quarries – large and small, disused and active. These holes in the ground are evocative reminders of an important and long-lived North Pennine industry. Another rock – whinstone – has also been quarried, but we'll discover more about that later in this book.

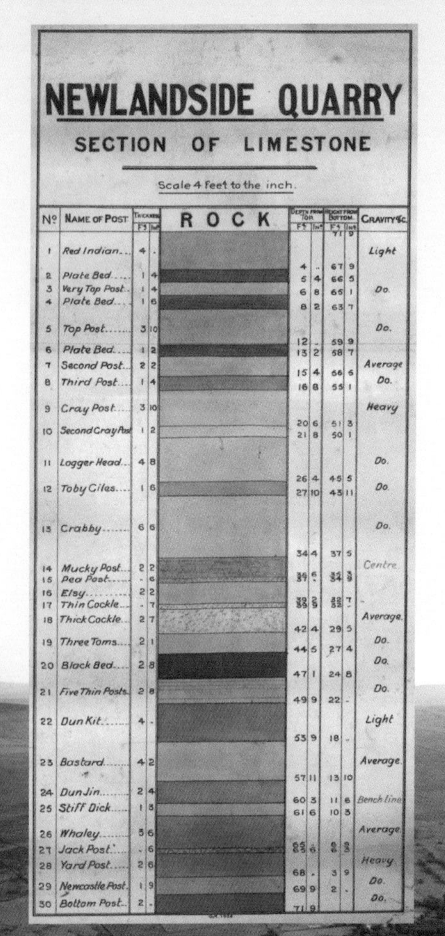

NEWLANDSIDE QUARRY

SECTION OF LIMESTONE

Scale 4 feet to the inch.

| No | Name of Post | Thickness Ft | Thickness In | Depth from Top Ft | Depth from Top In | Height from Bottom Ft | Height from Bottom In | Gravity &c. |
|---|---|---|---|---|---|---|---|---|
| | | | | | | 71 | 9 | |
| 1 | Red Indian | 4 | . | 4 | .. | 67 | 9 | Light |
| 2 | Plate Bed | 1 | 4 | 5 | 4 | 66 | 5 | |
| 3 | Very Top Post | 1 | 4 | 6 | 8 | 65 | 1 | Do. |
| 4 | Plate Bed | 1 | 6 | 8 | 2 | 63 | 7 | |
| 5 | Top Post | 3 | 10 | 12 | - | 59 | 9 | Do. |
| 6 | Plate Bed | 1 | 2 | 13 | 2 | 58 | 7 | |
| 7 | Second Post | 2 | 2 | 15 | 4 | 56 | 5 | Average |
| 8 | Third Post | 1 | 4 | 16 | 8 | 55 | 1 | Do. |
| 9 | Cray Post | 3 | 10 | 20 | 6 | 51 | 3 | Heavy |
| 10 | Second Cray Post | 1 | 2 | 21 | 8 | 50 | 1 | |
| 11 | Logger Head | 4 | 8 | 26 | 4 | 45 | 5 | Do. |
| 12 | Toby Giles | 1 | 6 | 27 | 10 | 43 | 11 | Do. |
| 13 | Crabby | 6 | 6 | 34 | 4 | 37 | 5 | Do. |
| 14 | Mucky Post | 2 | 2 | 36 | 6 | 35 | 3 | Centre |
| 15 | Pea Post | - | 6 | 37 | - | 34 | 9 | |
| 16 | Elsy | 2 | 2 | 39 | 2 | 32 | 7 | |
| 17 | Thin Cockle | - | 7 | 39 | 9 | 32 | - | - |
| 18 | Thick Cockle | 2 | 7 | 42 | 4 | 29 | 5 | Average. |
| 19 | Three Toms | 2 | 1 | 44 | 5 | 27 | 4 | Do. |
| 20 | Black Bed | 2 | 8 | 47 | 1 | 24 | 8 | Do. |
| 21 | Five Thin Posts | 2 | 8 | 49 | 9 | 22 | - | Do. |
| 22 | Dun Kit | 4 | - | 53 | 9 | 18 | - | Light |
| 23 | Bastard | 4 | 2 | 57 | 11 | 13 | 10 | Average. |
| 24 | Dun Jin | 2 | 4 | 60 | 3 | 11 | 6 | Bench line |
| 25 | Stiff Dick | 1 | 3 | 61 | 6 | 10 | 3 | |
| 26 | Whaley | 3 | 6 | 65 | - | 6 | 9 | Average. |
| 27 | Jack Post | . | 6 | 65 | 6 | 6 | 3 | |
| 28 | Yard Post | 2 | 6 | 68 | - | 3 | 9 | Heavy. |
| 29 | Newcastle Post | 1 | 9 | 69 | 9 | 2 | - | Do. |
| 30 | Bottom Post | 2 | - | 71 | 9 | | | Do. |

## Small and local

The North Pennines is dotted with hundreds of small disused quarries. Look out for shallow overgrown workings in fields and moors, and where crags have been cut into. For centuries, small quarries like these supplied local needs – sandstone for nearby walls and buildings, or limestone for making lime.

## Limestone for industry

The 20m-thick Great Limestone is the most quarried rock layer in the North Pennines. Weardale has a long history of working this limestone, which has been used as a flux in iron and steel making, and for making lime, cement, aggregate and roadstone. Limestone quarrying took off with the coming of the railways in the mid 1800s and flourished until the mid 1900s. Around 100 years ago, 1,200 men worked in the quarries around Frosterley and Stanhope, producing 20,000 tons of limestone a week.

◀ *Section through the Great Limestone in Newlandside Quarry near Stanhope, showing the individually named layers or 'posts'*

*An aerial view of Heights Quarry in Weardale* ▶

▲ Late 19th century or early 20th century quarrymen in the Parson Byers limestone quarry near Stanhope

## Still working

Limestone quarrying continues today at a handful of quarries. The Great Limestone is still worked in Weardale, but the industry today is highly mechanised and safety conscious – a far cry from the days of hard physical labour and hand tools. It also employs far fewer workers. In 1914 Ashes Quarry in Stanhope employed 200 men to produce 136,000 tons of rock a year; today, fewer than 20 men at Heights Quarry near Westgate produce over three times as much.

## Sandstone quarrying

For centuries the local sandstone has been quarried for use as a building stone. In the 18th and 19th centuries a large amount was used to line the shafts and tunnels of the area's lead mines. A type of very hard flinty sandstone, known as ganister, was also quarried for use as furnace linings. Several quarries still produce sandstone for paving, roofing, walling and building.

▲ Wildlife-rich wet areas in Ashes Quarry, Stanhope

## Back to nature

Many disused quarries are slowly being reclaimed by nature, creating rich habitats for wildlife. The rockfaces, spoil heaps and wet areas support many plants and animals and are particularly important for mosses, liverworts, lichens and invertebrates.

*Quarrying thin slabs of sandstone by hand in Ladycross Quarry* ▶

## WHERE TO SEE

▲ Old quarry in the Great Limestone, Bollihope Common

**Bollihope Common [NY 985 353]**
There are disused quarries in the Great Limestone on Bollihope Common, beside the road between Stanhope and Middleton-in-Teesdale.

**Ashes Quarry [NY 998 395]**
Good sections through the Great Limestone can be viewed from footpaths through this large disused quarry in Stanhope.

**Ladycross Quarry [NY 953 551]**
By arrangement it is possible to visit this historic quarry in Slaley Forest, where thin sandstone slabs have been hand quarried and worked for at least 300 years.

# Building and burning

The area's Carboniferous limestone and sandstone not only form the underlying structure of much of the North Pennine landscape, but also add distinctive character through the way they have been used over the centuries. The area's dry stone walls, buildings and settlements are all as much a part of this landscape as the fells and dales.

## Dry stone walls

Like much of upland northern England, this area is criss-crossed by dry stone walls, many of which date back 200 years or more. They were built of the nearest available stone, usually sandstone. This was either quarried nearby or gathered as clearance stones from surrounding fields. In some southern parts of the North Pennines, where limestone is the main rock type, the walls are built of limestone.

▲ *Houses of local sandstone in Blanchland*

## Building stone

North Pennine buildings and settlements derive much of their character from the Carboniferous sandstone of which they are made. These sandstones vary in grain size, colour and how they split. Some can be split very thinly and have traditionally been used as roofing slabs. Sandstone was also made into setts to pave the streets of places like Alston.

◄ *Repairing a roof in Blanchland with thin sandstone slabs from Ladycross Quarry*

▼ *Dry stone walls made of sandstone near Ramshaw*

## Moors and millstones

In the past, some thick layers of gritty sandstone were used to make millstones. There are millstone quarry sites, possibly dating back to the 18th century, high on craggy hillsides in Teesdale and Weardale. One site gave Millstone Rigg near Stanhope its name.

▲ *Millstones at Carr Crags, Upper Teesdale*

## Limestone to lime

Farmers have long recognised that lime-rich soils support good grassland, and for centuries have improved acid upland soils by spreading lime on the fields. Lime was made by burning coal and limestone in limekilns, many of which can still be seen dotted around the North Pennines. Lime was also used to make traditional lime mortar, and was produced for industrial purposes in large commercial limekilns.

▲ *Limekiln near Knock*

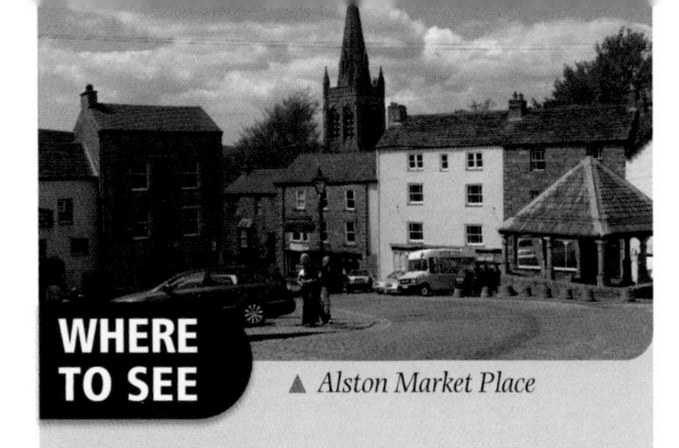

▲ *Alston Market Place*

## WHERE TO SEE

- **Alston [NY 718 465]**
  In Alston, the buildings and even the road are made of local sandstone.
- **Forest Head limekilns [NY 585 577]**
  An impressive set of commercial kilns, last used in the 1930s, lies next to a large disused limestone quarry and an old railway line.
- **Blanchland [NY 966 504]**
  The buildings and most of the roofs of this historic village are made of local sandstone.
- **Limekiln near Allenheads [NY 845 466]**
  A fine double-arched kiln lies just outside Allenheads, beside the road to Coalcleugh.

## Fossils and fonts

One local limestone has been prized as an ornamental stone since at least the 12th century. Known as Frosterley Marble, this layer within the Great Limestone is full of white fossil corals. There are fonts of Frosterley Marble in several Weardale churches, and it can be found in Durham Cathedral, York Minster and even Mumbai Cathedral!

*The Frosterley Marble font in St Michael and All Angels' Church, Frosterley* ▶

## Useful names

Quarrymen and miners gave names to many of the local rock layers, and some of these provide clues to how they were used in the past. Two sandstone layers, known as the Grindstone Sill and the Firestone Sill, were used to make grindstones and hearth stones. A 'sill' was a quarryman's term for a flat-lying layer of rock.

# Swamp forest

The Carboniferous deltas built up above sea level to form vast swampy areas, on which tropical rainforests flourished. The North Pennines would have been covered in lush swampy forest, which was periodically drowned when sea level rose. We can see relics of this today – in the area's coal seams and fossil trees and plants that can be found in many local sandstone layers.

## Swamp life

The Carboniferous forests would have been as dense and lush as today's Amazon rainforest, but they would have looked and sounded very different. Not only would the trees have been unrecognisable to us, but there would have been no birds or mammals. They had not yet evolved; even dinosaurs would not appear for another 100 million years! But the forests were not silent. Giant dragonflies flitted amongst the branches, cockroaches and scorpions scuttled on the forest floor, and large amphibians lived beside pools and rivers.

▲ *Fir clubmoss which grows today in rocky parts of the North Pennines*

## Giant moss

These forests contained some of the earliest large land plants. Trees up to 30m high towered above giant ferns and horsetails. These trees were unrelated to modern trees and included the ancestors of clubmosses. Modern clubmosses are small mountain plants, only a few centimetres high. The most advanced trees were primitive ancestors of modern conifers.

▲ *Fossil root in Maize Beck, showing the root surface texture*

▼ *A Carboniferous tropical swamp*

▲ *Thin coal seam exposed in Ladycross Quarry near Slaley*

## Black forest

Dead trees and plants built up as peat on the swampy forest floor. Over time this became compressed under layers of sediment, eventually hardening into coal. A one-metre-thick coal seam represents about 10m of dead vegetation and up to 10,000 years of forest growth. The soils in which the forests grew have been fossilised into layers known as seatearths, which may lie just beneath coal seams. Sometimes these fossil soils are preserved as pale, flinty sandstone, known locally as ganister.

## Cast in stone

The most common plant fossils in the North Pennines are of tree roots and bark. These formed when roots, tree trunks and branches were buried by delta sands which hardened around them. The wood rotted away, leaving spaces which became filled with sand; this then hardened into perfect sandstone casts.

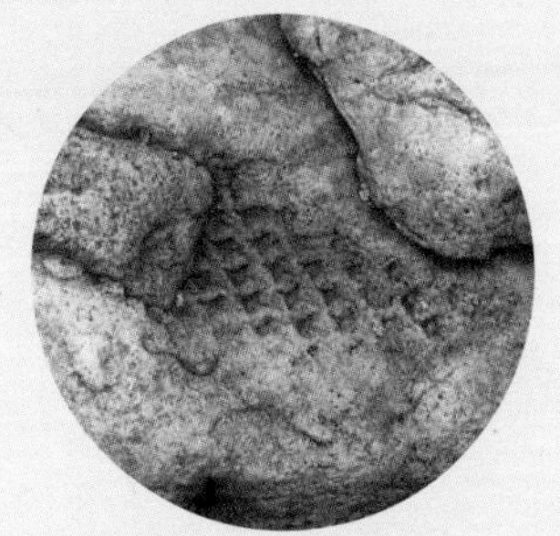

▲ *Small fossil branch in Swindale Beck, showing diamond-patterned bark texture*

## Coal mining

Most North Pennine coal seams are thin and were only worked on a small scale – look out for lines of small pits on hillsides. However, a few areas do contain thick, good quality seams. In the 19th and early 20th centuries the Tindale Fells area had many collieries. Until the 1920s, coal was also mined at Stublick near Langley and at Woodland in the south-east of the area. More recently, opencast extraction took place at Plenmeller in the late 1980s and early 1990s.

▼ *Stublick Colliery near Langley*

▲ *Stanhope fossil tree*

## WHERE TO SEE

**Stanhope fossil tree [NY 996 392]**
This is a superb example of *Sigillaria*, a giant ancestor of modern clubmosses. It was found in 1915 in a quarry at Edmundbyers Cross, north of Stanhope.

Look out for plant fossils in riverbeds, stone walls and stony tracks all over the North Pennines.

**Stublick Colliery [NY 833 604]**
The well-preserved 19th century buildings and chimneys of this colliery are very obvious at Langley.

**Gairs Colliery [NY 584 552]**
This colliery on the slopes of Tarnmonath Fell worked until the 1930s. You can see the remains of a railway, spoil heaps, buildings and a blocked-up mine entrance.

You can see small-scale coal workings as lines of small pits across the hillsides at Blagill in the Nent Valley and also near Coalcleugh.

# Molten rock

About 295 million years ago, just after the end of the Carboniferous Period, a momentous geological event took place beneath what is now the North Pennines. Stretching and splitting of the Earth's crust caused molten rock to rise up from deep within the Earth. This cooled underground to form the Whin Sill, now exposed at the surface as one of our most dramatic landscape features.

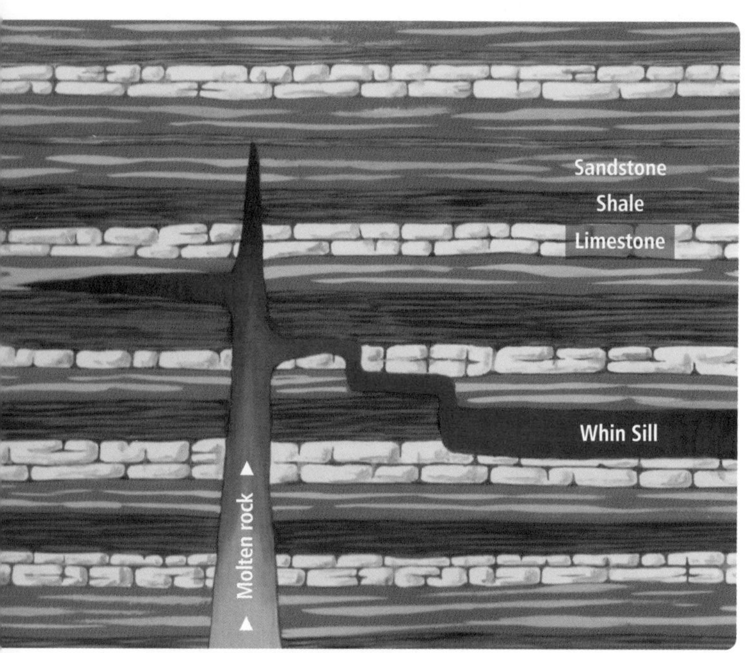


▲ *How the Whin Sill formed*

## Runny rock

This molten rock, or magma, was at a temperature of over 1,000°C. It didn't reach the surface but was forced sideways between layers of Carboniferous limestone, sandstone and shale. It cooled and solidified into a vast near-horizontal sheet, up to 90m thick, of hard, dark rock called dolerite, known locally as whinstone. The magma also formed a thin sheet of dolerite above the main Whin Sill, known as the Little Whin Sill. After millions of years of erosion these sills are now exposed at the Earth's surface in several places.

## Whin Sill dykes

Several vertical wall-like bodies of dolerite, known as dykes, cut across the North Pennines and may represent feeder systems for the Whin Sill. These include the Haydon Bridge Dyke in the north and the Hett Dyke near Hamsterley and Eggleston.

◀ *The Hett Dyke is a hard rib of dolerite, seen here forming the skyline north of Eggleston. The 'nicks' are glacial meltwater channels from the last ice age*

## Crystal kaleidoscope

▲ *A microscopic view of the Whin Sill, viewed under polarised light (actual size 3mm across)*

If you look closely at a piece of fresh dolerite, you can see that it is made up of a mass of small interlocking crystals – mainly pyroxene (dark) and feldspar (white). These minerals crystallized out of the magma as it cooled. When cut into very thin slices, only about 0.03mm thick, and viewed under a polarising microscope, the Whin Sill is strikingly beautiful, with jewel-like colours and intricate crystal shapes.

▲ *Holwick Scars, Upper Teesdale*

## Cracking columns

The Whin Sill probably took around 50 years to cool underground, from molten rock to solid dolerite. During the final stages of cooling it contracted, producing vertical cracks along which the rock breaks into rough columns – a feature known as columnar jointing. You can clearly see these cracks and columns in Whin Sill cliffs and quarry faces.

▼ *Dramatic cliffs of Whin Sill dolerite at High Cup Nick*

## Sugar Limestone

When the Whin Sill was molten, it had a huge effect on surrounding rocks. In Upper Teesdale, adjacent limestone was baked and altered to a white, crumbly, crystalline rock. Although this is known as Sugar Limestone, it is in fact a marble. To a geologist, marble is a limestone that has been altered by heat or pressure, or both, in the Earth's crust. This unusual rock and its soils support the unique 'Teesdale Assemblage' of arctic-alpine plants, including the beautiful spring gentian.

▼ *Sugar Limestone near Cow Green Reservoir, Moor House–Upper Teesdale National Nature Reserve*

▲ *Low Force*

## WHERE TO SEE

- **High Force [NY 880 284], Low Force [NY 903 280] and Cauldron Snout [NY 814 286]**
  These attractive waterfalls in Upper Teesdale are all formed by the Whin Sill.
- **Holwick Scars [NY 904 269]**
  Holwick is one of the most accessible places to see dramatic crags of the Whin Sill.
- **High Cup Nick [NY 746 262]**
  On the North Pennine escarpment the Whin Sill forms a spectacular ring of cliffs.

The Whin Sill is also spectacularly exposed outside the North Pennines. It is a natural rampart for Hadrian's Wall in the Northumberland National Park, and forms the Farne Islands and some dramatic coastline in the Northumberland Coast AONB.

# Working the whinstone

The hardness of Whin Sill dolerite has not only created dramatic landscape features, but also makes it a readily accessible and commercially valuable material. There is a long history of whinstone quarrying in the North Pennines and this industry continues to this day.

## Streets of whinstone

In the 19th and early 20th centuries, as well as being crushed to make roadstone, whinstone was also shaped by hand into block-shaped 'setts' and kerbstones. These were used to pave the streets of Britain's growing industrial cities.

▲ *Middleton Quarry in the early 20th century, showing piles of setts*

▼ *A disused whinstone quarry in Teesdale*

▲ *Working the whinstone at Force Garth Quarry*

## On the road

Today, the quarrying tradition continues at Force Garth Quarry in Upper Teesdale, where the Whin Sill is worked to make roadstone. Dolerite is particularly suitable for surfacing roads because of the way it is made up of an interlocking mass of different minerals. These minerals wear down at different rates, meaning that the surface of the rock never becomes completely smooth and always retains a slight roughness – just what is needed for a road.

## What's in a name?

The Whin Sill took its name from northern quarrymen's terms – 'whin' was a hard dark rock and a 'sill' was any flat-lying layer of rock. When 19th-century geologists worked out the true nature and origin of the Whin Sill, the word 'sill' was adopted for all similar bodies of once-molten rock worldwide. The Whin Sill is therefore famous for being the original sill of geological science.

*Force Garth Quarry, Upper Teesdale* ►

▲ *Building made of dolerite in Holwick, Upper Teesdale*

## Hard rock

Dolerite was rarely used as a building stone as it is extremely hard and difficult to work, especially when compared with the local sandstone which is a good building stone. However, in places like Holwick in Upper Teesdale, which are very close to outcrops of the Whin Sill, a few buildings are made of dolerite.

▲ *Greenfoot Quarry*

## WHERE TO SEE

**Greenfoot Quarry [NY 985 391]**
This disused and flooded quarry near Stanhope produced crushed rock from the Little Whin Sill (a thinner sheet of dolerite which lies above the main Whin Sill).

**Holwick [NY 906 269]**
The unusual use of dolerite as a building stone can be seen in buildings and walls at Holwick. Look out for blocks of very hard dark stone.

**Force Garth Quarry [NY 873 282]**
Best viewed from the Pennine Way on the opposite side of the River Tees, this active quarry works dolerite for roadstone.

Several disused whinstone quarries, mainly in Upper Teesdale, can be seen from public roads.

# Hot water and minerals

The North Pennines is world-famous for its remarkable deposits of lead ore and other minerals, which make up part of the Northern Pennine Orefield. They formed when minerals crystallized out of warm mineral-rich solutions circulating deep underground. These deposits have been mined for centuries and have yielded many superb specimens.

## Granite and hot water

North Pennine mineral deposits are believed to have formed about 290 million years ago. Mineral-rich waters warmed by heat from the buried Weardale Granite flowed through cracks deep underground. As the fluids cooled, their dissolved minerals crystallized within the cracks, forming mineral veins. Sometimes, the fluids reacted with limestone on the sides of the cracks, altering the rock and forming mineral deposits known as flats.

▲ *Minerals in a flat at Rogerley Mine, Frosterley, including shiny grey galena and green fluorite*

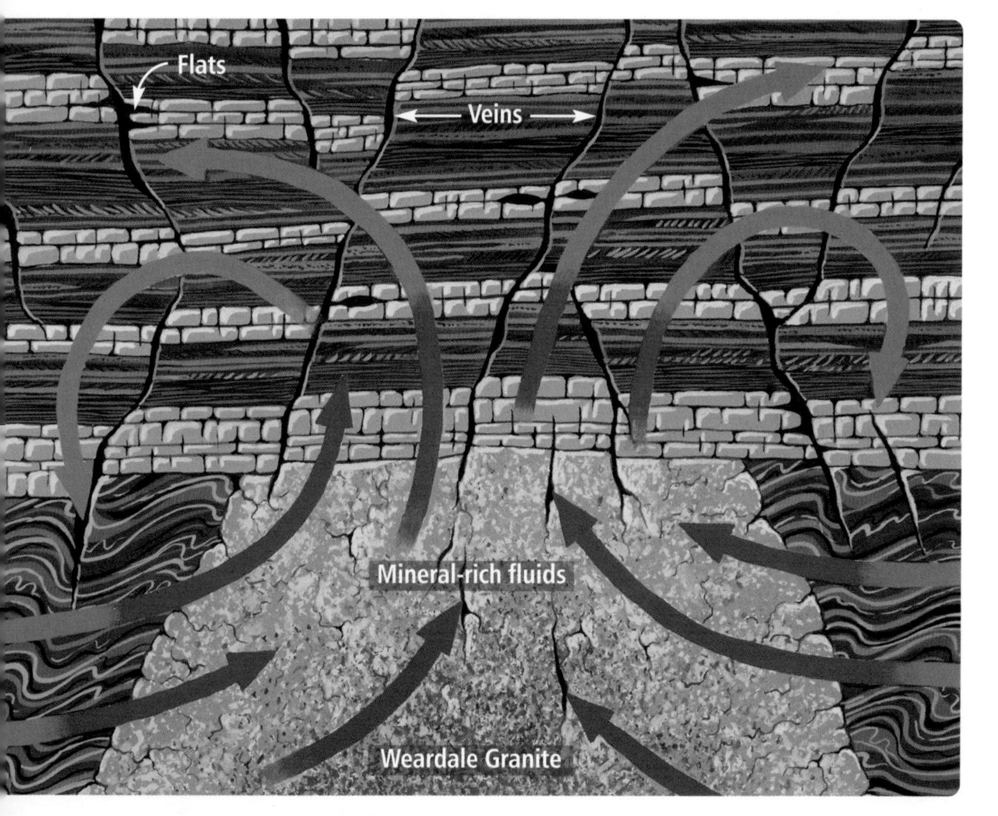


◄ *How North Pennine mineral veins and flats formed*

## What's in a vein?

North Pennine mineral veins vary from a few millimetres to many metres wide. A typical vein consists of bands of minerals parallel to the sides of the veins. Minerals such as fluorite, baryte and quartz commonly make up most of the vein, with smaller amounts of ore minerals like galena (lead ore) and sphalerite (zinc ore). Adjacent flats tend to be rich in iron ores.

## In the zone

The mineral deposits are distributed in a zoned pattern – rather like an irregular bull's-eye. These zones are thought to reflect the cooling of the mineral-rich fluids as they flowed away from the centre of the orefield. The central zone, which includes much of Alston Moor, Weardale, and the East Allen and Derwent valleys, contains abundant fluorite, which crystallized between 100 and 200°C. The outer zone contains barium minerals such as baryte and witherite, rather than fluorite. Baryte crystallized at temperatures as low as 50°C.

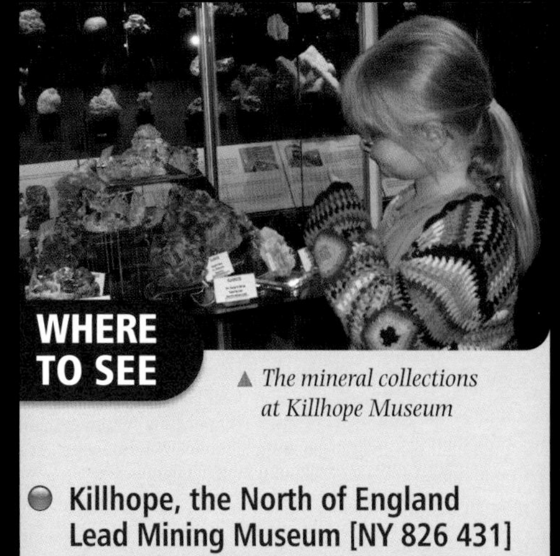

**WHERE TO SEE**

▲ *The mineral collections at Killhope Museum*

**Killhope, the North of England Lead Mining Museum [NY 826 431]**

Visit Killhope Museum to see some superb specimens of the area's minerals and find out more about how they formed.

## Spectacular minerals

▲ *Yellow fluorite crystal on white baryte crystals from Hilton Mine, Scordale*

North Pennine mines have yielded many superb mineral specimens, including some of the world's best examples of fluorite. These can be found in most of the world's major mineralogical collections, along with fine specimens of witherite, barytocalcite and alstonite. These three minerals were first recognised in the North Pennines and are very rare elsewhere in the world.

▲ *Galena (large, dull, grey crystal) and sphalerite (small, dark, shiny crystals) from Smallcleugh Mine, Nenthead*

## Research and discovery

The North Pennines has a long history of research on mineral deposits. Ideas and concepts developed here by miners and mine managers, and later by geologists and mineralogists, have been fundamental to the understanding of mineral deposits around the world.

▼ *Fluorite crystals from Rogerley Mine, Frosterley*

# Lead mining landscapes

North Pennine mineral deposits were the foundation of the area's economy for many centuries. Most important was lead mining, which had its heyday in the 18th and 19th centuries. With its legacy of mine sites, smelt mills and settlements, lead mining has had a profound effect on North Pennine landscapes and communities.

## Millennia of mining

Lead mining in the North Pennines has a very long history; the Romans almost certainly mined lead ore here. Over the centuries the mines developed from shallow pits to large opencast and underground workings. Lead mining boomed in the 18th and 19th centuries when this area was one of the world's leading producers of lead. But by the late 1800s the industry was in decline, leaving a legacy of shafts, spoil heaps, mining settlements and deep trenches known as hushes.

◄ *Lead miners working in Smallcleugh Mine, Nenthead, 1897*

## Beyond the mine

There is much more to the story of North Pennine lead mining than just the mines themselves. Material from the mine had to be washed and sorted on 'washing floors' to separate valuable ore from waste rock. The ore was then smelted at smelt mills to extract molten lead, which was shaped into ingots and taken to the wharves and warehouses of Tyneside. Until the early 19th century, lead ore and finished lead were carried by packhorses along 'carriers' ways'. All these processes left their imprint on the landscape.

▲ *The washing floor at Allenheads Mine in the 1870s*

▼ *Chimney of engine house at Sikehead Mine above Ramshaw, near Blanchland*

▲ *Lead mine entrance and spoil heaps at Nenthead*

▲ *The miner-farmer landscape around Allenheads*

▲ *Smelt mill flue above Ramshaw, near Blanchland*

## Silver lining

Most galena (lead ore) contains small amounts of silver. Every ton of North Pennine lead contained a few ounces of silver – a significant by-product of lead mining. In medieval times the silver may have been more valuable than the lead, so some early mining may have been for silver.

## Miner-farmers

Many lead-mining families made ends meet by working small farms. This way of life created a distinctive 'miner-farmer' landscape in parts of the North Pennines; the pattern of isolated farm buildings and small fields can still clearly be seen today in many North Pennine dales.

## Lead-loving plants

The ground at many lead-mining sites is toxic to plants and almost bare of vegetation – even though the mines were abandoned over 100 years ago. But some plants – metallophytes – tolerate or even thrive on soils with high levels of lead and zinc. These include alpine pennycress, mountain pansy and spring sandwort, also known as leadwort.

*Spring sandwort* ▶

▲ *Killhope Museum*

## WHERE TO SEE

- **Killhope, the North of England Lead Mining Museum [NY 826 431]**
  Here you can take a mine tour, have a go on the washing floor and find out how the miners lived and worked.
- **Allenheads [NY 859 453]**
  Find out about the area's mining heritage upstairs in the Blacksmith's Shop.

Look out for mine entrances, spoil heaps, hushes and chimneys all over the North Pennines.

# More than lead

Lead mining was the most important industry in the North Pennines for centuries, but many other minerals have been mined too. Iron, zinc and copper ores, as well as non-metallic minerals like fluorite and baryte, have all been worked in the past. These too have all left their imprint on the landscape.

## Iron out

Mining for iron ore is a rather forgotten industry of the North Pennines. Like lead mining, it has a long history – probably dating back 2,000 years. After flourishing in medieval times, iron mining boomed again in the 19th century. Iron ores from Weardale supplied many local blast furnaces like those at Tow Law and Consett. Iron mining had largely ceased by the early 20th century.

▲ *Iron-rich rocks in front of the Governor and Company's Level in the Middlehope Valley, a mine which was worked for lead and iron ores*

## Black jack

Known as 'black jack' by the miners, sphalerite (zinc ore) was once dumped on lead mine spoil heaps. However, once zinc became a valuable metal in the 19th century, old lead mines were reopened and spoil heaps were reworked for zinc ore. During the Second World War, old spoil heaps at Nenthead were reworked for zinc. One of zinc's many uses is in galvanising steel.

▼ *The remains of Groverake Mine near Rookhope, the last working fluorspar mine in the North Pennines*

▲ *A new mine tunnel being constructed in Allenheads in 1970, in an unsuccessful search for fluorspar*

## Fabulous fluorite

Fluorite (known commercially as fluorspar) is one of the iconic minerals of the North Pennines. In the 19th century it was a waste product of lead mining, although the miners appreciated its beauty and collected the best crystals to decorate their homes and make decorative 'spar boxes'.

In the 20th century fluorite became valuable as a flux in the iron and steel industries, and as a source of fluorine chemicals. As demand grew, old lead mines and spoil heaps were reworked. The last fluorspar mine closed at Rookhope in 1999.

▲ *Industrial buildings in Allenheads which date from the 1970s attempt to mine fluorspar*

## Heavy spar

Baryte (known commercially as barytes) is one of several barium minerals found in the North Pennines. It is a heavy white mineral, called 'heavy spar' by the lead miners. Barytes mining started in the 1880s for use in the growing paint and chemical industries. It is also used in paper-making, for making oil drilling fluid and in radiation shields. The last barytes mine, at Closehouse in Lunedale, closed in 2001.

▲ *West Rigg Opencut, Westgate*

## WHERE TO SEE

- **West Rigg Opencut [NY 911 392]**
  This is an old working for iron ore in flats on either side of the Slitt Vein.
- **Slitt Wood [NY 905 400]**
  There are piles of rusty-brown iron ore amongst the mine workings at Middlehope, at the northern end of Slitt Wood.
- **Groverake [NY 896 441]**
  The headgear of the last North Pennine fluorspar mine can be seen from the road between Rookhope and Allenheads.
- **Great Rundale, Dufton [NY 714 276]**
  Mines at the head of this valley changed from lead to barytes production in the 1880s. The spoil heaps were reworked for barytes in the 1980s.
- **Allenheads [NY 860 454]**
  Modern industrial buildings in the middle of Allenheads date from an attempt to mine fluorspar in the 1970s.

▲ *Large opencast barytes working at Closehouse Mine in Lunedale*

# Desert plains

At the time the North Pennine mineral deposits were forming, Britain was drifting north of the equator. Between 290 and 210 million years ago we were in latitudes that are today occupied by the Sahara and the Arabian Gulf. We were in the Permian and Triassic periods and Britain was part of a vast continent known as Pangaea. Northern England lay just north of the equator and was once again a hot and arid place.

## Dunes and rivers

Wind-blown sand was sculpted into huge dunes, and desert lakes periodically dried up leaving salt flats. Scree accumulated at the foot of rapidly eroding hills, and vast river systems spread across the plains, depositing red silts and sands. The sands from the deserts and rivers hardened into the red sandstones of the Eden Valley. These include the Permian Penrith Sandstone and the Triassic St Bees Sandstone. The salts from the desert lakes are preserved as layers of the minerals anhydrite and gypsum. Gypsum is mined today near Dufton for use in plasterboard and plaster.

▲ *Old quarry in St Bees Sandstone at Croglin*

## Clues to the past

Evidence of these ancient desert environments is preserved in the rocks. The sand grains in the Penrith Sandstone are spherical, of a type known to geologists as 'millet seed' grains. Grains like this form in deserts where they are rounded by relentless wind-blasting. The slopes of large dunes are also preserved in the sandstone. By looking at these shapes geologists can tell that the wind in this ancient desert blew from the east or south-east.

▲ *Close-up of Penrith Sandstone, showing rounded sand grains (view is 2cm across)*

▼ *How the North Pennine escarpment might have looked 250 million years ago*

▲ Brockram in a barn near Kirkby Stephen

## Life in the desert

There would have been life in this harsh environment. Red sandstones of the same age elsewhere in the Eden Valley and in southern Scotland contain fossil footprints and rare bones, showing that large reptiles roamed desert Britain.

## Brockram

Brockram is the local name for a Permian breccia – a rock made of coarse, angular fragments of older rocks. Brockram formed from scree and flash flood deposits and is made of fragments eroded off nearby hills. It consists mainly of fragments of pale grey Carboniferous limestone set in red sandy material.

## Stone for building

The red Permian and Triassic sandstones are easy to work and shape into blocks for building. They were often used in a decorative way around arches and windows. The sandstone is relatively soft – as can be seen by the eroded and rounded shapes of stones in old walls and houses.

▼ Local red sandstone used in St Lawrence's Church, Kirkland

▲ Red sandstone houses in Dufton

## WHERE TO SEE

The best places to see these rocks are in the villages along the escarpment foot where many of the houses are built of red Permian and Triassic sandstone.

### Kirkby Stephen [NY 776 087]

Brockram can be seen in the walls of the lane leading down to Frank's Bridge; the large field barn a little further up the river also contains many blocks of brockram in its walls.

### Dufton Ghyll [NY 692 248]

Hidden in this wooded gorge are cliff faces of red St Bees Sandstone which were quarried in the past – you can still see the tool marks in places.

### Penrith and Eden Museum [NY 515 303]

Here you can see a fossil reptile footprint in Penrith Sandstone, from a quarry near Penrith. It is around 20cm wide by 30cm long.

# Mind the gap

From over 200 million years ago to the last ice age, we have little evidence for what happened in the North Pennines; very few rocks survive from this period. Millions of years of landscapes, rocks and creatures have been lost forever – almost without trace. But this doesn't mean nothing happened. By studying rocks from other parts of Britain, geologists can reconstruct what happened here.

## Dinosaurs and ammonites

▲ *Ammonite from Whitby on the Yorkshire coast*

In the Jurassic Period – 200 to 145 million years ago – shallow seas covered much of Britain, depositing clay and limestone. Giant reptiles and ammonites lived in the sea, dinosaurs roamed the land and pterosaurs soared overhead. The North Pennines was probably above sea level for most of the time. Any rocks that did form here have long since worn away. However, Jurassic rocks and fossils of some of the amazing creatures from this time can be seen on the Yorkshire coast.

## Chalk seas

The Cretaceous Period, from 145 to 65 million years ago, was also a time of warm seas. Hundreds of metres of chalk – a soft white limestone made of tiny shells – accumulated. Chalk forms the White Cliffs of Dover and the rolling Yorkshire Wolds. The North Pennines rose above the seas for some of the time but was probably once covered by chalk that has long since disappeared.

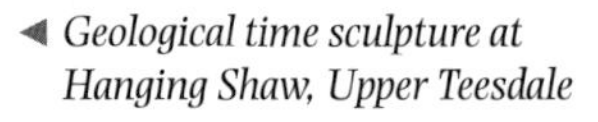
◄ *Geological time sculpture at Hanging Shaw, Upper Teesdale*

▲ *Chalk cliffs near Dover*

▲ *The Cleveland Dyke, here forming a rocky rib at Tynehead*

## Birth of an ocean

About 60 million years ago the North Atlantic Ocean started to open. The Earth's crust split and molten rock erupted out of volcanoes in what is now western Scotland. Cracks underground radiated out from volcanic areas, providing conduits for molten rock. This solidified to form vertical sheets of rock or dykes. One of these is the Cleveland Dyke, which stretches 400km from Mull to the Yorkshire coast. It cuts across the North Pennines, linking our landscape with momentous geological events that were happening far away.

## Uplift and erosion

After the opening of the North Atlantic, more Earth movements affected the North Pennines. These were related to the formation of the Alps far to the south. As Africa bulldozed into Europe, the effects of this collision rippled northwards. The North Pennines was uplifted along large faults, which still define the northern, southern and western edges of the North Pennines. After a long period of weathering and erosion, in warm and humid conditions, the climate cooled dramatically about 2.6 million years ago – we had entered a period of ice ages!

▲ *Exposed patch of weathered dolerite at Holwick*

### WHERE TO SEE

**Holwick Scars [NY 901 270]**
This outcrop of rotten Whin Sill dolerite may have been weathered in warm and humid conditions during Palaeogene and Neogene times.

**Tynehead [NY 757 365]**
The road and stream cross the Cleveland Dyke which forms a grass-covered rib.

▼ *Volcanic activity continues today in Iceland, as the North Atlantic gradually widens*

# Ice age

Britain drifted north to its present position, and about 2.6 million years ago world climate cooled dramatically, heralding the start of a series of ice ages. At the height of the last major ice age, about 20,000 years ago, northern Britain was blanketed in vast ice sheets and would have looked like Antarctica or central Greenland today. The North Pennine landscape we know today owes much to the action of ice and meltwater.

## In and out of the freezer

Over the past 2.6 million years the North Pennines has been covered in ice many times. Long periods of arctic conditions known as 'glacials' alternated with warmer 'interglacials' when the climate was similar or even warmer than today. We are currently in an interglacial, but one day – thousands of years in the future – ice will return, creeping down from the north to once again blanket the North Pennines.

## Scratching the surface

During the last glacial period the North Pennines lay under a vast ice sheet up to a kilometre thick. The ice was full of boulders, gravel and sand and acted like a giant sheet of icy sandpaper. It streamed roughly eastwards, although at different times and in different places, the ice flow directions varied. By scouring and scraping the landscape, the ice widened and modified pre-existing valleys and smoothed the hills.

▼ *This view of Antarctica shows how the North Pennines might have looked about 20,000 years ago*

▲ *The sweeping shape of High Cup Gill was sculpted by ice and meltwater*

▲ *Beldon Cleugh, west of Blanchland, is an impressive sinuous meltwater channel*

## Under pressure

Powerful meltwaters flowing beneath the ice carved drainage channels through solid rock. Sometimes the water flowed uphill under immense pressure, cutting channels that are unrelated to modern drainage.

## Asymmetrical valleys

Valleys that run north-south, like the Allen Valleys, lay at right angles to the direction of ice movement. As the ice streamed eastwards it scoured the eastern valley sides but dumped glacial debris on the western sides. This created valleys in which one side is smooth and covered in glacial debris, whereas the other displays terrace features.

## Tracking the ice

Boulders that have been carried by ice and dumped away from their bedrock source are known as erratics. They are scattered around the North Pennines. Boulders carried from the Lake District and southern Scotland around the edges of the North Pennines allow geologists to track ice flow directions.

## Tills and hills

The ice sheet smeared the landscape with a mixture of clay, sand and boulders known as glacial till. As ice flowed down valleys it moulded till at its base into whaleback-shaped mounds which are streamlined in the direction of ice movement. Known as drumlins, these can be seen today as rolling grassy hillocks in Upper Teesdale and the Eden Valley.

▲ *Drumlin at Harwood, Upper Teesdale*

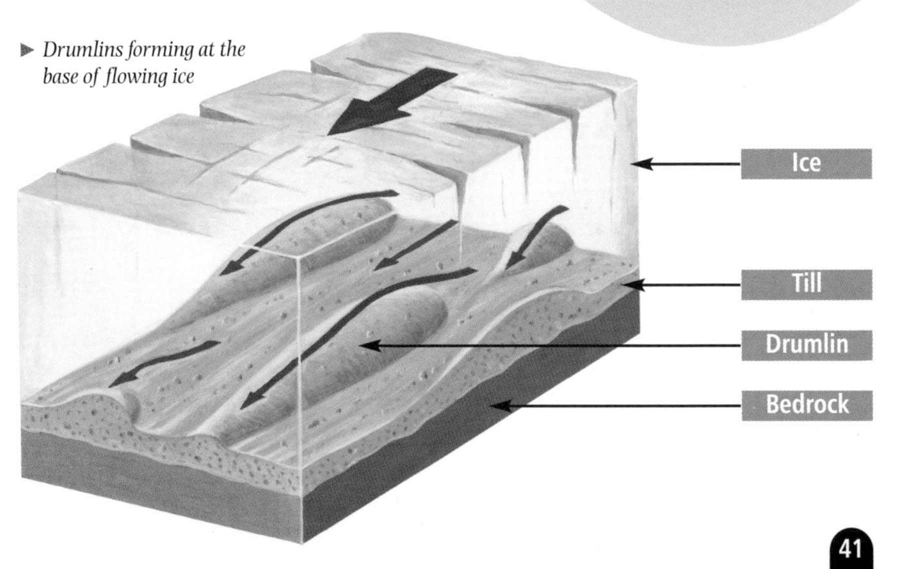


► *Drumlins forming at the base of flowing ice*

▲ *The Bullman Hills*

## WHERE TO SEE

**Bullman Hills [NY 705 371]**
These limestone erratics are so large that they form small hills on the slopes of Cross Fell.

**Cow Green [NY 820 315]**
Limestone erratics dot the moor north of the reservoir road. They form small mounds distinguished by greener grass than the surrounding moorland.

**Holwick [NY 910 270]**
The rolling grassy hills around Holwick are drumlins, formed under ice streaming down Teesdale.

**High Cup Gill [NY 740 255]**
This dramatic valley is an example of a valley that at least partly owes its U-shape to the action of ice and meltwater.

**Beldon Cleugh [NY 913 504]**
This meltwater channel winds up and over a ridge, showing that it formed under pressure beneath the ice.

# Scarred land

About 15,000 years ago the climate became warmer and wetter and the arctic conditions released their grip on the North Pennines. The ice began to melt, amidst torrential meltwaters, leaving a scarred landscape. The shapes of the fells and dales would have been recognisable to us, but they would have looked very different – this was a desolate land of bare rock and glacial debris.

▲ *A mound of glacial sand and gravel near Knock Pike which has been partly quarried away*

## Torrents of water

Torrential meltwaters poured out of the margins of the retreating ice sheet, depositing sheets of sand and gravel. Material carried by meltwaters running around the edges of the ice, or in crevasses and tunnels within the ice, was left behind as mounds and ridges of sand and gravel.

## Kettle holes

Large chunks of ice, which became stranded and covered by glacial debris as the ice retreated, melted where they lay to form hollows known as kettle holes. These filled with water to become ponds and lakes. Talkin Tarn is the largest and best example of a kettle hole in the area.

◀ *Talkin Tarn near Brampton*

▼ *This view of the Canadian Arctic shows how the North Pennines might have looked towards the end of the ice age*

▲ *The scar left by a large landslip on the slopes of High Cup Gill*

## Slip sliding away

Melting of the ice left behind an unstable landscape. There was no vegetation to bind the sediments so loose sand, gravel and shattered rock were easily eroded by rivers and streams. Saturated glacial debris and rock slumped and slid downhill, often where weak shale layers failed. Many of the steep-sided valleys along the North Pennine escarpment show the scoop-shaped scars of landslips with debris piles below.

▲ *The infilled, pre-glacial channel of the River Tees at Cauldron Snout (marked by line)*

## Changing course

Some rivers altered course because of the way the landscape had been modified by ice. At Cauldron Snout the old pre-glacial course of the River Tees was plugged by till so the river was forced to cut a new route through the Whin Sill. It has therefore taken it around 15,000 years to form the channel through which it now cascades.

▲ *Talkin Tarn*

## WHERE TO SEE

- **Talkin Tarn [NY 545 587]**
  Talkin Tarn is a kettle hole, left behind when a large piece of stranded ice melted.
- **Knock Pike [NY 685 288]**
  North of Knock Pike are mounds of glacial sand and gravel, one of which has been worked as a gravel pit beside the road.
- **Cauldron Snout [NY 814 286]**
  At the foot of the waterfall there is a pre-glacial river channel now infilled by till.

## Patterns in stone

Even after the ice had gone, the North Pennines would have been cold and barren. Repeated freezing and thawing created blockfields of frost-shattered rock, and caused heaving of the ground surface. This led to the formation of unusual accumulations of stones, including stripes and polygons, on the high slopes and plateaus. Many of these are relict features, but on the very highest hilltops they are probably still active.

*Blockfield on the slopes of Cross Fell* ▶

# Life on the edge

Once the ice had melted, the North Pennines was not barren for long. Plants and animals colonised the bare land, bringing it to life. Eventually, people arrived too, heralding a new chapter in the evolution of the landscape.

## Green revolution

As the ice retreated, the first colonisers – mosses, grasses, arctic plants and shrubs – started to green the bare and barren landscape. Soon the North Pennines would have been dotted with small plants, similar to those of arctic regions today. Some arctic-alpine plants, such as the spring gentian, survive at Cow Green in Upper Teesdale. This landscape was inhabited by tundra animals like arctic fox and reindeer.

◀ *Spring gentians near Cow Green Reservoir*

▼ *Tundra vegetation, including dwarf shrubs, growing today in Greenland*

◀ *Juniper woodland survives today in Upper Teesdale and is easily viewed from the Pennine Way near High Force*

## Wildwood

In time, trees arrived and spread over the landscape. Pollen preserved in peat shows that juniper and birch were followed by hazel, Scots pine, oak and elm. Patchy open woodland developed across the North Pennines, covering all but the highest hilltops. With the trees came forest animals – wolf, brown bear, lynx, red deer, wild boar and aurochs, extinct prehistoric cattle.

▲ *Bronze Age ring cairn on Birkside Fell, which contained a burial urn. The area has also yielded Mesolithic flint tools*

## Early settlers

We know very little about the people who shared this landscape with the aurochs and wolf, but we do know that hunter-gatherers were living here from about 10,000 years ago. Flint tools from their hunting camps have been found buried beneath peat. About 8,000 years ago people began felling trees to create clearings, probably to help with hunting animals like deer – and so began the long story of human influence on the landscape.

▲ Sphagnum *mosses*

▲ *Eroding peat near Great Rundale Tarn*

## A blanket of peat

About 7,500 years ago rainfall increased. A wetter climate, perhaps combined with tree clearance by people, heralded the demise of woodland and the start of peat formation on the uplands. Peat forms from an accumulation of *Sphagnum* mosses and other plants. These don't fully decay in waterlogged conditions and over thousands of years they build up into a layer of peat, which can be several metres deep.

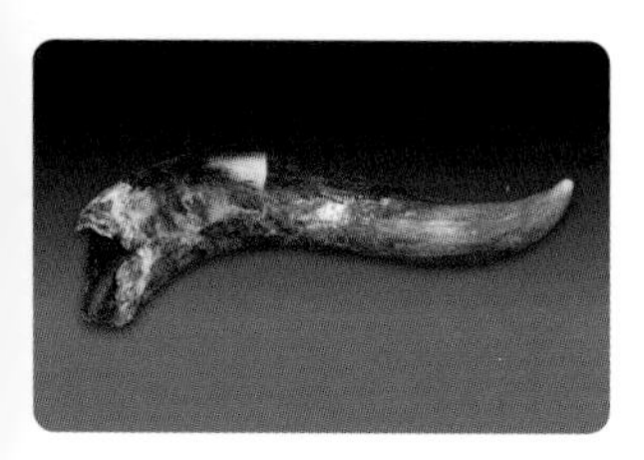

▲ *An aurochs horn from Ireshope Moor in Upper Weardale*

## In the bog

Preserved in the blanket bogs of the North Pennines are glimpses into our prehistoric past. Ancient pollen trapped within the peat layers can tell us how vegetation has changed over thousands of years. The peat also preserves relics of the ancient forests that once covered the area – look out for bleached tree stumps and birch twigs. Aurochs horns, including one dated at 4,000 years old, have also been found buried under peat in Teesdale and Weardale.

▲ *Ancient tree stump at Smiddy Shaw Reservoir*

## WHERE TO SEE

**Smiddy Shaw Reservoir [NZ 046 461]**

Ancient Scots pine stumps are emerging out of the peat at the edge of the reservoir.

Peat covers much of the uplands and you can see sections through it in many places, especially along the edges of tracks, streams and reservoirs.

**Cow Green [NY 817 297]**

A unique assemblage of arctic-alpine plants, including the spring gentian, grows on the lime-rich soil of this part of the Moor House–Upper Teesdale National Nature Reserve.

▲ *The aurochs stood about two metres high at the shoulder, and would have towered above modern cattle*

# Today's landscape

Today's landscape is the product of millions of years of Earth processes and a few thousand years of human activity. And it is still evolving on a range of different timescales – both geological and human. Some changes take place incredibly slowly and imperceptibly, whereas others are rapid and dramatic. One thing we can always be sure of – the landscape is never still…

## The long view

The processes that have shaped the landscape throughout Earth history continue today. We are still moving with the Earth's tectonic plates and in millions of years Britain will have journeyed to a different position on the globe. The rocks we see today may erode and be recycled to form new rocks containing fossils of creatures yet to evolve.

## Worn away

The fells are continually being eroded as rocks tumble off crags or are worn down by water. Eroded material is washed away by streams and deposited downstream. Flooded rivers can change the landscape in just a few hours by washing away riverbanks in some places and creating new deposits elsewhere. In limestone country, the rock gradually dissolves so that sinkholes and caves get larger over time.

▲ *A view of Holwick in Upper Teesdale showing modern settlements and the stone walls of post-medieval fields overlying evidence of much older field systems*

## A working landscape

For thousands of years people have worked this landscape. Agriculture in particular has played a major role through woodland clearance, enclosure by stone walls, liming and drainage. The area also bears witness to centuries of mining and quarrying, and the settlements that grew up around these industries. This is still a working landscape, where farming, grouse moor management, forestry and quarrying continue to play a part in its evolution.

▲ *North Pennine waterfalls gradually move upstream as they erode the underlying layers of limestone, sandstone and shale*

▲ *Rockfall in a small gorge cut through limestone, Maize Beck near High Cup Nick*

▲ *The River Derwent in flood through Blanchland in July 2009*

▲ *Repairing dry stone walls at Knarsdale, South Tyne Valley*

## Looking after the landscape

Today, conservation is also playing an important role. Work to block drains on the moors is helping to restore our blanket bogs. New native woodland is being planted, and there is work to conserve upland hay meadows and other special habitats. Projects to look after historic buildings and teach traditional skills like dry stone walling are also helping to care for our built heritage.

## Changing times

Our climate has changed over millions of years – as we've seen in this book. Changes on shorter timescales are not well understood and it is hard to make accurate predictions about future climate and what impact this will have. Whatever happens, we know that the North Pennines will continue to change. By looking after our special landscapes we will give the natural features, habitats, wildlife and communities that are part of them the best chance of thriving in the future.

◀ *Spreading species-rich green hay in Upper Teesdale, as part of a project to restore and enhance upland hay meadows*

▼ *Blocking drains or 'grips' on peatland near Killhope, Upper Weardale, using plugs of peat*

# Useful books and maps

We hope you have enjoyed finding out about the geology and landscape of the North Pennines. This book provides only a flavour of the area's story and we hope it has whetted your appetite to find out more. Below is a selection of useful publications and maps.

## Publications

Beale, S. and Dodd, M. (eds) 2008. *Exploring Lakeland Rocks and Landscapes.* The Cumberland Geological Society.

Bowden, A. 2008. *History in the Landscape: The rocks, landscape and minerals of Weardale.* The Weardale Society.

Bulman, R. 2004. *Introduction to the Geology of Alston Moor.* North Pennines Heritage Trust.

Dodd, M. (ed) 1992. *Lakeland Rocks and Landscape: A Field Guide.* The Cumberland Geological Society.

Forbes, I., Young, B., Crossley, C. and Hehir, L. 2003. *Lead Mining Landscapes of the North Pennines Area of Outstanding Natural Beauty.* Durham County Council.

Johnson, G.A.L. (ed) 1995. *Robson's Geology of North East England.* Transactions of the Natural History Society of Northumbria. Vol. 56, Part 5.

North Pennines AONB Partnership. 2006. *Wheels to the Wild Cycle Route: Discovering Geology and Landscape in the North Pennines.*

North Pennines AONB Partnership. 2010. *North Pennines Area of Outstanding Natural Beauty and European Geopark. A Geodiversity Audit. Revised March 2010.* Available as a pdf from **www.northpennines.org.uk** (as are other geological leaflets and geotrails).

Scrutton, C.T. (ed) 2004. *Northumbrian Rocks and Landscape: A Field Guide (2nd edition).* The Yorkshire Geological Society.

Skipsey, E., Webb, B. and Young, B. 2002. *Rocks and Landscape of Alston Moor.* Cumbria RIGS and East Cumbria Countryside Project. Available as a pdf from **www.eccp.org.uk/images/great-days-out/rocks-landscape5.pdf**

Stone, P. and six others. 2010. *British Regional Geology: Northern England (5th edition).* Keyworth, Nottingham: British Geological Survey.

Symes, R.F. and Young, B. 2008. *Minerals of Northern England.* National Museums Scotland.

## Geological maps and memoirs

The following British Geological Survey maps cover the North Pennines (several have associated detailed memoirs or shorter explanations):

**1:50,000 scale:** Brampton (18), Hexham (19), Newcastle (20), Penrith (24), Alston (25), Wolsingham (26), Appleby (30), Brough-under-Stainmore (31), Barnard Castle (32), Kirkby Stephen (40).
**1:25,000 scale:** Cross Fell Inlier, Middleton-in-Teesdale.

Several of the maps are covered by the classic publication: Dunham, K.C. 1990. *Geology of the Northern Pennine Orefield Vol. 1 Tyne to Stainmore.* Economic memoir of the British Geological Survey. 2nd edition.

For details of British Geological Survey publications contact the Sales Desk (0115 936 3241, sales@bgs.ac.uk, **www.bgs.ac.uk**).